The Fossil
Evidence
for Human
Evolution

The University
of Chicago Press
Chicago
and London

Third Edition
Revised & Enlarged
by Bernard G.
Campbell

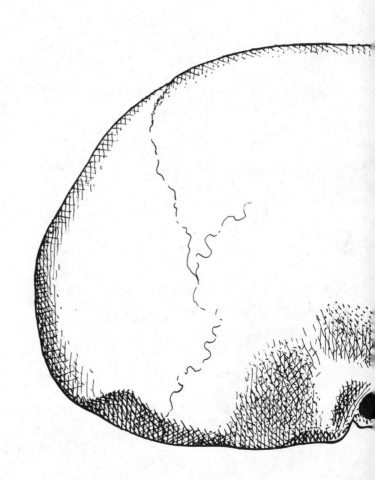

The Fossil Evidence for Human Evolution

An Introduction to the Study of Paleoanthropology

W. E. LeGros Clark

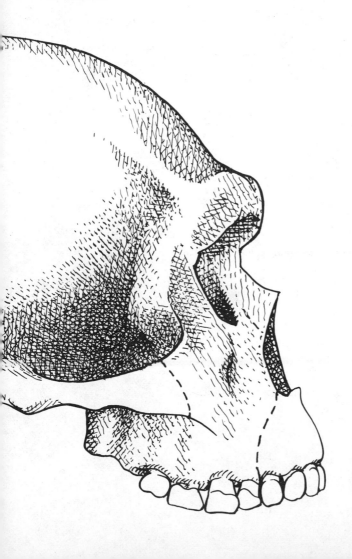

Bernard G. Campbell is adjunct professor of
anthropology, University of California at Los
Angeles. He is the author of *Humankind
Emerging*; *Human Evolution*; and *Sexual
Selection and the Descent of Man*.

The late W. E. LeGros Clark was professor of
anatomy at Oxford University and the author of
The Antecedents of Man; *The Tissues of the
Body*; and *Chant of Pleasant Exploration*.

The University of Chicago Press, Chicago 60637
The University of Chicago Press, Ltd., London

Library of Congress Cataloging in Publication Data

Clark, Wilfrid Edward Le Gros, Sir, 1895-1971.
    The fossil evidence for human evolution.

    Bibliography: p.
    Includes index.
    1. Fossil man.   2. Human evolution.   I. Campbell,
Bernard Grant.   II. Title.   [DNLM: 1. Evolution.
2. Primates.   3. Paleontology.   4. Phylogeny.
GN282 C596f]
GN282.C56 1978   569'.9   78-52 9
ISBN 0-226-10937-2
ISBN 0-226-10938-0 pbk.

Contents

Contents

Illustrations

Preface to the
Third Edition

In undertaking to prepare a new edition of this book I have become deeply aware of my great debt to Sir Wilfrid Le Gros Clark and to his brilliant essay on the fossil evidence for human evolution. The principles set out so clearly in the first edition (1955) stand today as they did then, as the central and lasting basis on which to interpret these fossils. The preface to the second edition also requires no modification: although the fossil evidence has doubled in volume since the second edition was published in 1964, we are still a long way from assessing its full significance and agreeing on its interpretation. Yet some general consensus does exist, and in this book a broad interpretation of the present evidence has seemed possible, though much remains to be done. In particular, the status of *Ramapithecus* is not yet secure. From this early phase of hominid evolution we need a great deal more fossil evidence, and it is in this early period that we can expect our science to develop particularly rapidly in the coming years.

It has been a pleasure to work on this new edition and to get to know its author somewhat better. If Sir Wilfrid were still alive, I hope he would be pleased to know that his book now has new life and that it is held in great esteem by students and teachers throughout the world.

Bernard G. Campbell

Preface to the
Second Edition

It seems desirable that this book should be prefaced by some
explanation of its purpose. It is not intended, of course, to
provide an account of all the fossil evidence bearing on hominid
evolution, or even the greater part of it. It aims only to present
some of the main sources of evidence in an abbreviated form and
to indicate such conclusions as may justifiably be drawn from
them. But it is concerned even more to examine critically the
logical basis of the arguments and inferences which from time to
time have been advanced on the basis of studies of fossil material
or of studies of the comparative anatomy of man and his nearest
living relatives. Not so many years ago, fossil remains of hominids
and anthropoid apes were very scanty; indeed, to those paleontologists who deal with vertebrate groups richly represented in the
fossil record they must have appeared almost ridiculously so. In
recent years they have been accumulating to a remarkable extent,
and because of their importance in the elucidation of the origin of
our own species, they naturally have a most compelling and
personal interest. It is the more necessary, therefore, that their
significance should be assessed objectively in accordance with the
well-established principles of paleontological science. Paleoanthropologists are now becoming more critical in their appraisal
of the available fossil evidence, and it is well that this should be
so, for it is important, in considering the possible significance of

new discoveries, to avoid some of the confusions and misunderstandings which have too often characterized discussions on human evolution in the past. Many of these misunderstandings have a terminological origin, and it has become more and more clear that they can be avoided in the future only if paleoanthropologists agree to the use of a common taxonomic nomenclature based on adequate definitions.

With the rapid accumulation of the fossil record of the Hominidae, much of the new material lately discovered has still to be described and compared in full detail. With such considerations in mind, it is quite apparent that some of the conclusions based on this paleontological evidence must necessarily be of a provisional nature. But it is right and proper to attempt an assessment of the available evidence from time to time, even if conclusions can be no more than provisional, for only by so doing is it possible to formulate working hypotheses which may be put to the test as more data come to hand.

Phylogenetic interpretations based on a fossil record which is still far from complete are, of course, meant to be no more than interpretations. They are offered for confirmation or modification as the record becomes more and more complete. Thus it is not claimed that the conclusions presented in this book are in any sense final. On the other hand, it is suggested that at least they accord reasonably well with the facts so far as these are at present available. And the facts which have in recent years become available now constitute a sequential record which, though by no means complete, does provide a substantial documentation of hominid evolution.

W. E. Le Gros Clark

One

Morphological and Phylogenetic Problems of Taxonomy in Relation to Hominid Evolution

It is well recognized by comparative anatomists that there is a biological relationship between man and the anthropoid apes, in the sense that, by a process of gradual diversification, both have been derived in the distant past from a presumed common ancestral stock. When, in the early days of evolutionary studies, this inference was first made, it was based almost entirely on the totality of the anatomical resemblances between men and apes. Similar degrees of resemblance between other groups of mammals had been accepted as evidence of an evolutionary relationship; and in some cases this relationship had been further demonstrated and confirmed by the discovery of fossil remains of extinct creatures of intermediate type. But at the time of the publication of Darwin's book *The Descent of Man* in 1871 the fossil record of human evolution was almost nonexistent, and the evidence for the relationship of men and apes—although it seemed convincing, even then, to many biologists—was only indirect. In more recent years, the fossil record bearing on human ancestry has accumulated to a remarkable extent. It is no longer meager in comparison with that of the evolution of some other mammalian groups, for it has provided much more direct and concrete evidence for the relationship of men and apes than was previously available. It is particularly noteworthy how closely some of these fossil types conform to intermediate stages of human evolution, which had been postulated and predicted on the basis of the indirect evidence of comparative anatomy. Discoveries of such fossil relics, indeed, provide a remarkable vindication of the well-established methods of comparative morphology that have been used to assess systematic affinities and phylogenetic relationships. As Watson (1951) has noted, in arriving at a natural classification "individual animals . . . are allotted to their groups by structural similarities and differences, determined by observations directed by ordinary morphological reasoning. And the methods of morphology are shown to be valid because they enable us to make verifiable predictions." The progress of paleontological discovery has led to the verification of many of the predictions (based on the study of the comparative anatomy of living species) regarding the phylogenetic relationships and evolutionary origins of different groups of Primates. In this sense, indeed, paleontology might almost be called an experimental science, that is, if "experiment" is defined (as it is in the *Oxford English Dictionary*) as "a procedure adopted for testing an hypothesis." Comparative anatomical studies of living forms may demonstrate a gradational series of morphological types, and such a series may suggest an actual temporal sequence of evolution. But the inference of a temporal sequence based on indirect evidence of this sort can be validated only by the direct evidence of paleontology.

Comparative Anatomy and Taxonomy

Before considering the fossil evidence for the zoological relationship between man and ape, a brief reference may be made to the indirect evidence of anatomical resemblance. In spite of superficial appearances, these resemblances are actually very close. The bony skeleton, for example, is constructed on the same general plan; and in the case of some of its elements it may hardly be possible to make a clear distinction without a careful scrutiny of morphological details or the comparison of biometric data. It is precisely for this reason, of course, that rather sharp contro-

versies have occasionally been aroused in the past over the identification of certain fossil fragments of a doubtful nature—
whether they should be referred to ape or man. The muscular
anatomy of men and apes is astonishingly alike, even down to
some of the smaller details of attachments of many of the
individual muscles. The similarity in the structure and disposition
of the visceral organs suggests that apes have a closer relationship
to men than they have to the lower Primates. The human brain
(though larger in relation to body size) in its morphology is little
more than a magnified model of the brain of an anthropoid ape;
indeed, there is no known element of the brain in the former that
is not also to be found developed to some degree in the latter. All
these facts, together with observations recording similar meta-
bolic processes, serological reactions, chromosome patterns,
blood groups, and so forth, are now well known. It is because of
such a striking complex of resemblances that in schemes of
zoological classification man and the anthropoid apes have for
many years been placed quite close, and in recent years the
tendency has been for a still closer approximation.

Even the preevolutionary biologists of the past classified man in
the Mammalia (they could hardly do otherwise), but they emphasized their conception of his apartness by placing him in a
separate order, or even in a separate subclass, of mammals. More
detailed and less subjective studies later placed him in the order
Primates, according him a family status (Hominidae) that is
equated with the Pongidae (anthropoid apes) and the Cercopithecidae (catarrhine monkeys). Today many authorities believe
that the detailed morphological resemblances between man and
the anthropoid apes are more accurately expressed by grouping
the Hominidae and Pongidae in a common superfamily, Hominoidea, and thus contrasting them both with the catarrhine
monkeys—Cercopithecoidea (Simpson 1945). The closer association in a zoological classification of man and apes is no doubt due
partly to the stricter application of taxonomic criteria, particularly as the result of the work of those authorities who are
specially qualified by practical experience to apply the principles
of vertebrate taxonomy to the Primates. But it is also due to the

discovery of the fossil remains of primitive hominids in which the morphological distinctions contrasting *Homo sapiens* with the Recent anthropoid apes are not nearly so obtrusive.

Notwithstanding the numerous anatomical resemblances between Recent men and Recent apes, there are, of course, pronounced differences, and in the past these have all been duly emphasized by one anatomist or another. In some cases they have certainly been overemphasized, partly (it seems) because the human anatomist tends by the nature of his studies to focus attention on the minutiae of morphology and is thus inclined to exaggerate their taxonomic significance. Perhaps, also, the very personal nature of the problem has led some authorities to lay more stress on differences between men and apes than they would lay on equivalent differences in other mammalian groups. The lists of morphological characters that from time to time have been put forward as evidence of man's "uniqueness" among mammals often give the strong impression that the authors were straining the evidence to the utmost limit (and sometimes a considerable way beyond it) in order to substantiate their thesis. It does not always seem to be fully realized that some of the unique features of *H. sapiens* on which the authors laid stress are merely distinctions at a generic or specific level, or that they represent little more than an extension of morphological trends that are quite apparent in other Primates. Nor was adequate account taken of the fact that similar claims for uniqueness might be equally valid for many other mammalian species. The giraffe, for example, possesses a number of anatomical characters that establish it as "unique" among all other living mammals; but its taxonomic position in the Giraffidae (equated with related families of the same infraorder, Pecora, such as the Cervidae and Bovidae) is accepted as a reasonable interpretation of its evolutionary status. Anatomically, *H. sapiens* is unique among mammals only in the sense that every mammalian species is in some features unique among mammals. The unique character of *H. sapiens* lies primarily in his behavior.

It is instructive to consider a sample of those anatomical differences that have been claimed to be so exclusively distinctive of the Hominidae as to demand an unusual degree of taxonomic isolation, and to see how these equate with differences found among other mammals that are accepted as forming natural groups in the phylogenetic sense. In *Homo* the large size of the brain relative to the body weight is certainly a feature that distinguishes this genus from all other Hominoidea, but it actually represents no more than an extension of the trend toward a progressive elaboration of the brain shown in the evolution of related Primates (and also of many other groups of eutherian mammals). In its morphological pattern, indeed, the human brain shows much less contrast with that of the large anthropoid apes than is found among the various families of an equivalent taxonomic category of Primates, the superfamily Lemuroidea; and even in its absolute size it shows less contrast than is to be found between the primitive and advanced genera of the single family Equidae. Several patterns of articulation between the bones of the skull have been claimed as distinctive of the Hominidae, but they also occur as variants (in some instances rather commonly) in the skulls of apes. In any case, however, the differences in pattern of articulation that are claimed to distinguish the Hominidae from the Pongidae are no more pronounced than those that occur between different groups of, say, the single family Cercopithecidae. The dentition of the Hominoidea, in spite of differences between the Hominidae and Pongidae such as the relative size and shape of the canines and first lower premolars, actually shows far closer similarities than are to be found in the several families of, for example, the Feloidea. The linear proportions of the limbs in *H. sapiens* show certain well-recognized differences from those of the Recent anthropoid apes, but again these differences are not more marked than they are in, say, the several families of the Muroidea. Even the form of the glans penis and the fact that the female is characterized by permanent, well-developed breasts have been adduced as evidence of "man's" uniqueness among mammals. But these may be no more than specific characters of *H. sapiens* and not characteristic features of the Hominidae as a whole (for, of course, we have no idea at all what the penis and the female breast may have been like in extinct types of man such as *Homo erectus*). These examples (which could well be multiplied) appear to indicate that anatomists who have laid such stress on differences of this sort as emphasizing the morphological "uniqueness" of *H. sapiens* have perhaps introduced a personal and subjective element into their assessment of the taxonomic status of their own species.

Man, *Homo sapiens*, and the Hominidae

To eliminate the subjective factor as far as possible in discussions on hominid evolution, it seems essential to avoid colloquial terms such as "man" and "human." In recent controversies on the taxonomic position of fossil hominids, as so frequently in the past, their too common use has been a very obvious source of confusion. The fact is that these terms may not properly be used as though they were equivalent to the zoological terms *Homo* and Hominidae or to the adjectival form "hominid" in the same way that "horse" can be substituted for "equid." In the latter case, although the primitive equid *Hyracotherium* is so markedly different superficially from the modern *Equus*, it can be called a "primitive horse" without real danger of misunderstanding, for the term "horse" (or "horse family") still remains elastic enough in the minds of most people to permit its extension to relatively remote ancestral forms. Similarly, the fossil Hominoidea of Miocene age may appropriately be called "primitive anthropoid apes," even though they had not acquired all the specialized features that are accepted as characteristic of the anthropoid apes of today (see p. 000). But the terms "man" and "human" have come to assume, by common usage, a much narrower and more rigid connotation, which for most of us (however we may try to persuade ourselves otherwise) also involves a real emotional element. There can be little doubt that if these colloquial terms were to be rigidly excluded in strictly scientific discussions of the evolutionary origin of *H. sapiens*, and only the terms proper to taxonomy employed, such problems could be approached on a much more objective plane. The confusion to which the loose use

of the term "man" may give rise is well illustrated by those
comparative studies in which the skeletal elements of a fossil
hominoid are compared with only the Recent anthropoid apes
and with only *H. sapiens* (or perhaps even with only one or two
racial varieties of *H. sapiens*) and the conclusion is then drawn
that in certain features the fossil agrees more closely with the
"anthropoid apes" than with "man." The fallacy of such a
statement is obvious, but it may nevertheless be very misleading
to the casual reader. Clearly, the authors of such studies are using
the term "man" as though it were equivalent to *H. sapiens*, or
perhaps only one racial variety of *H. sapiens*. But the term "man"
should also be taken to include extinct types, such as Neandertal
man and the various representatives of *H. erectus*. These should
be taken into consideration and also, of course, the extinct types
of anthropoid ape. The fallacy of extrapolating from single
species or genera to larger taxonomic groups is not uncommon in
the literature of paleoanthropology.

Homo sapiens is one of the terminal products of an evolutionary radiation that also led to the development of other types that
have now become extinct, all of which are included in a natural
group, the family Hominidae. If this familial term is to be used as
it is in the definition of equivalent mammalian groups (i.e., with
due reference to the fundamental concepts of evolution and to the
taxonomic designation of other equivalent radiations), it must
include not only *H. sapiens* but all those representatives of the
evolutionary lineage or sequence that finally led to the development of this species (and of collateral lines) from the time when
the sequence first became segregated from the evolutionary sequence of the Pongidae. Similarly, the family Pongidae logically
includes not only the Recent anthropoid apes (gorilla, chimpanzee, orang, and gibbon) but also the fossil and extinct related
types to which this particular line of evolutionary development
gave rise after it had become clearly segregated from the common
ancestral stock that also gave rise to the Cercopithecidae. It is of
the utmost importance that those taking part in discussions on
phylogenetic sequences and zoological classification should
recognize the implications of these taxonomic terms and make

proper use of them. So far as the Hominoidea are concerned, they are illustrated diagrammatically in figure 1. From this diagram it will be seen that (as will be discussed in later chapters) three genera of the Hominidae are here provisionally recognized: *Homo* —represented by the three species *H. sapiens, H. erectus* (an extinct type, previously regarded as a separate genus *Pithecanthropus*, which existed in the Far East and Africa during Early and Middle Pleistocene times) and the recently discovered *H. habilis*—and second, *Australopithecus*, a primitive hominid whose fossilized remains have been recovered in great quantity from travertine deposits in South Africa and, more recently, from stratified deposits in East and northeast Africa. It still remains a matter of argument whether *Australopithecus* should be split into two or more genera. Third, *Ramapithecus* comprises a poorly represented group of ground-living species that are closely related to the Pongidae but on present evidence are believed to have given rise to *Australopithecus*.

The phylogenetic relationships of hominid genera to one another and to the anthropoid ape family (Pongidae) indicated in the diagram are to be regarded as provisional interpretations based on the fossil evidence so far available. But the main intention of the diagram here is to make it clear that the Hominidae include all those genera and species that represent the earlier and later developments of a single evolutionary radiation, the latter being distinguished from the other radiation of the Hominoidea—the Pongidae—by a clear divergence of evolutionary trends. Similarly, the genus *Homo* or *Australopithecus* includes those species that represent in each case a common line of evolutionary development diverging from other species of the Hominidae. In other words, these terms are taken to connote a "vertical" classification of the Hominoidea, in the sense that they are indicative of separate main and subsidiary evolutionary sequences, ascending and diversifying from common basal stocks.[1] It is particularly urged that this simple and now generally accepted scheme of classification of the Hominoidea should be adhered to and that other terms invented from time to time by

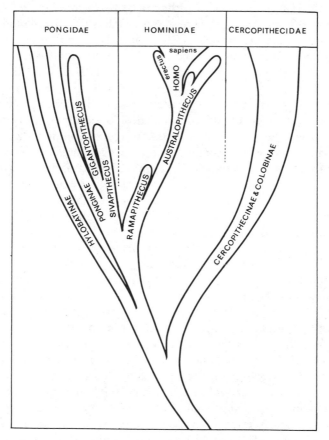

Fig. 1. Diagram illustrating the probable phylogenetic relationship of the known genera of the Hominidae, and the relationship of the Hominidae to the Pongidae. This diagram is also intended to emphasize that, as paleontological records in general clearly establish, representatives of an earlier genus may survive and persist long after other representatives of the *same* genus have given rise to a later genus. In other words, an ancestral genus may for some time actually be contemporary with its derivative genus.

individual authors (such as Euhominids, Prehominids, Prehominians, Archanthropinae, Presapiens, and so forth) should be strenuously avoided. Insofar as none of these terms has ever been properly defined, they remain entirely obscure, and, insofar as they remain obscure, their continued use is bound to lead to confusion.

One of the most important desiderata of a workable classification is that it should meet with a general (even if only provisional) adoption for common usage. Taxonomic systems usually involve a compromise of vertical and horizontal classifications, the former coming to predominate more and more as the fossil evidence of phylogenetic lines gradually accumulates; while this evidence remains incomplete, no two students of a particular mammalian group are likely to agree on all points of their classification. But they can, and should, agree to use a common system even if they do not accept all its implications, provided that the system represents at least a reasonable approximation to probable phylogenetic relationships and that it is broadly conceived in its basic plan. In this work we make use of Simons's classification of the Primates (1972), not because it is the only possible classification, or on the assumption that it represents the last word on phylogenetic relationships (which no system of classification can do until the fossil record approximates a complete record). I use it because (1) it is based on recognized authority and experience, (2) it has the merit of simplicity, (3) it appears to reflect reasonably closely such phylogenetic relationships as can be inferred from the evidence at hand (and as far as this can be done in any system of classification without seriously affecting convenience of reference; but see note 1), and (4) it has been provisionally accepted and recognized by other authoritative workers in the same field. It is well to emphasize this last consideration, for only by the provisional acceptance of a common scheme of classification (with common and well-understood taxonomic terms) is it possible for zoologists or anthropologists to engage in discussions on problems of Primate evolution without misrepresentations and misunderstandings.

One of the most common sources of confusion in discussions on hominid evolution is the tacit assumption by some writers that a large brain is an essential character of the Hominidae; this example is referred to here because it illustrates rather forcefully the misuse of taxonomic terms. Now it is, of course, obvious that one of the outstanding characteristics of *H. sapiens* is the relatively large size of the brain. But some of those anatomists who lay most stress on this distinction also maintain that the Hominidae (i.e., the family comprising the evolutionary radiation leading, inter alia, to *H. sapiens*) separated from the other radiations of the Primates at a very remote geological period—at least by the Early Miocene. (See chart on p. 186.) On the other hand (as we shall later see), there is no evidence to show that the precursors of *H. sapiens* acquired a brain size approaching modern dimensions before the Early Pleistocene—certainly not before the Upper Pliocene. Consequently, there must presumably have been a very lengthy period in the hominid sequence of evolution during which the brain still retained a size approximating that of the modern large apes. In other words, although a large brain may be accepted as one of the diagnostic characters of the species *H. sapiens*, it is not a valid criterion of the family Hominidae. In a well-reasoned essay, Oakley (1951) has suggested that the term "man" (and presumably "human" as well) should be reserved for those later representatives of the hominid sequence of evolution who had reached a level of intelligence indicated by their capacity to fabricate implements of some sort. Man, that is to say, is essentially a tool-making creature. If this definition is accepted, then the earlier small-brained representatives of the Hominidae who had not yet developed this capacity may conveniently be referred to "the prehuman phase of hominid evolution."

The assumption that a brain of large dimensions is a distinctive feature of the Hominidae as contrasted with the Pongidae is certainly in part the result of a terminological confusion resulting from the loose usage of the terms "man" and "human" as though they were equivalent to "hominid." But it has also been argued (again mainly on the basis of comparative anatomical studies of living forms) that the primary factor that led to the evolutionary divergence of the Hominidae from the Pongidae was the rapid enlargement of the brain and that other distinctive characters, such as those related to posture and gait, were secondary developments. However, apart from the fact that functional considerations make such a speculative hypothesis unlikely or, indeed, impossible, this argument is rendered untenable by the paleontological data now available.

It has been necessary to divaricate on the significance of the taxonomic terms used in discussions on hominid evolution, because much fruitless argument has evidently been expended for want of careful definitions, and it may be well to reemphasize briefly one or two of the major points discussed. The purpose of zoological classification, as Simpson (1945) has pointed out, is primarily a matter of practical convenience; but the basis of a zoological taxonomic system should be phylogenetic, and the criteria for the definition of taxonomic terms should, as far as is consistent with practical convenience, be consonant with the evidence of phylogeny. It is for this reason, of course, that zoological classifications must necessarily be provisional and tentative in the absence of a substantial paleontological record, and they need constant revision as this record becomes more and more complete. In the absence of any fossil remains, the family Equidae would presumably be defined (as Cuvier, indeed, defined his equivalent of this family) on the basis of the existing species of *Equus*—that is to say, by characters such as the single complete digit on each foot and the complicated pattern of the cheek teeth. But with the accession of fossil material the definition of Equidae has been widely extended to include such primitive representatives of the equid sequence of evolution as *Hyracotherium* and *Mesohippus*. In defining the Hominidae and contrasting this family with the Pongidae, phylogeny must also be taken into account. In other words, a satisfactory definition is to be obtained only by a consideration of the fundamental factors of structural evolution that characterized the initial segregation of the Hominidae and the Pongidae from a common ancestral stock, and of the divergent trends the two evolutionary sequences followed in their later development. The former consideration is no doubt of more direct importance in the identification of fossil remains of early hominoids whose taxonomic status may not be immediately clear, for in such primitive types some of the later evolutionary trends will not yet have manifested themselves to any marked degree. Those anatomical features that probably have no more than a specific value for the definition of *H. sapiens* (such as the combination of a large brain with a vertical forehead, small teeth, reduced jaws, presence of a chin, and so forth) must obviously be avoided in the definition of the Hominidae.

Though in any system of classification the lower taxonomic categories, such as genera and species, may be defined in more or less static terms (at least insofar as they concern divergent rather than successional types), it is impracticable to draw up any comprehensive definition of larger categories, such as families and orders, except on the basis of evolutionary trends. The latter may be inferred indirectly by a consideration of the various end-products of evolution in each group, but they can be ultimately demonstrated only by paleontological sequences. For example, the order Primates is particularly difficult to define by reference to fixed characters, mainly because, as a group, it is not distinguished by any gross forms of adaptive specialization possessed by all its members in common. The evolutionary progress of the Primates, as Simpson (1950*a*) has well said, has been in the direction of greater adaptability rather than of greater adaptation. Thus the order can be defined only by reference to the prevailing evolutionary trends that have distinguished it from other groups—such as the progressive development of a large and complicated brain, the elaboration of the visual apparatus and a corresponding reduction of the olfactory apparatus, the abbreviation of the facial skeleton, the tendency toward the elimination of the third incisor tooth and of one or two premolars, the preservation of a relatively simple pattern of the molar teeth, the replacement of sharp claws (falculae) by flattened nails (ungulae), the retention of pentadactyl limbs with an accentuation of the mobility of the digits, and so forth (Clark 1954). Not all Primates (even those that exist today) have developed all these characters or completed their development to the same degree. And still less so, of course, was this the case with the earliest Primates of the Eocene period.

The Early Differentiation of
the Hominidae and Pongidae

As we have already noted, the expansion of the brain to the large dimensions characteristic of *H. sapiens* was evidently a relatively late phenomenon in the sequence of hominid evolution. What, then, are the basically distinctive characters whose development initiated the evolutionary radiation of the Hominidae in the first stages of its segregation from the Pongidae and that might therefore be expected to be of importance for the differentiation of the *earlier* representatives of these two families? From a consideration of the details of comparative anatomy of living forms and from the evidence now available from fossil hominoids, it appears reasonably certain (and, indeed, is agreed by authorities who hold widely differing views on a number of phylogenetic details) that the most important single factor in the evolutionary emergence of the Hominidae as a separate and independent line of development was related to the specialized functions of erect bipedal locomotion (Washburn 1950). Herein the Hominidae showed a most marked evolutionary divergence from the Pongidae, for in the latter the strongly contrasting brachiating mode of locomotion was developed. In the one case the lower limb increased in length relative to the trunk length and to the length of the upper limb; in the other it was the upper limb that increased in relative length. In other words, in the relative growth of the limbs the Pongidae and the Hominidae have followed two opposing allometric trends. In the Hominidae the bony elements of the foot and the knee joint became modified in shape and proportions to permit the structural stability required for bipedal progression; in the Pongidae the mobility of these parts was enhanced for specialized prehensile functions. In the Hominidae the pelvic skeleton underwent far-reaching changes directly related to the erect posture; in the Pongidae it retained the general shape and proportions found in most lower Primates. These divergent modifications of the limbs and pelvis are related to very different modes of life in the two families. They involve more than just those proportional differences in linear dimensions that can be determined by overall measurements and about which so much detailed and accurate information has been accumulated by the patient studies of Schultz (1931, 1936, 1950*b*), for they are also accompanied by marked structural divergences in muscular anatomy (as Straus [1949] has emphasized). The total morphological pattern of the limbs and pelvis in the Hominidae, where they are known, thus presents a criterion by which these are distinguished rather abruptly from the known representatives of the Pongidae. If, as now appears probable, this distinction was the factor responsible for the primary segregation of these two evolutionary radiations, it provides a most important clue for assessing the true taxonomic position of their early representatives.

Another criterion of special importance in paleontology is the total morphological pattern of the dentition. In all the Recent Pongidae the primitive features of the Primate dentition have been modified by the widening of the incisor series, the replacement of a forwardly converging molar-premolar series by parallel (or slightly divergent) tooth rows, the accentuation of powerful overlapping canines (with a pronounced sexual dimorphism), and the development of a more strongly sectorial character of the anterior lower premolar. It may be noted here that a sectorial type of lower premolar, which is characteristic of all Recent anthropoid apes, is a predominantly unicuspid tooth, the main cusp having anteriorly a cutting edge that shears against the upper canine (see fig. 34*B1*). In all the later known Hominidae the incisors remain small, the molar-premolar series converge anteriorly in a rounded arcade, the canines have undergone a relative reduction in size and do not overlap to any marked degree (and not at all after the early stages of attrition, except in some individuals of the Pleistocene genus, *Homo erectus*), and the anterior lower premolar is of a bicuspid, nonsectorial form. These differences in the dentition may have been secondary to those of the limbs and pelvis, but the paleontological evidence indicates that they also became manifested relatively early in the evolutionary history of the two families. In the absence of other evidence, recognition of *Ramapithecus* depends at present on features of the dentition.

Total Morphological Pattern

Reference was made earlier to the "total morphological pattern" presented by limb structure or the dentition. It seems desirable to stress this concept of pattern rather strongly because the assessment of the phylogenetic and taxonomic status of fossil hominoid remains must be based not on the comparison of individual characters one by one in isolation, but on a consideration of the *total pattern* they present in combination. Vertebrate taxonomists are, of course, well accustomed to taking account of groups of characters in their assessments of the zoological status of an animal, and they are quite conversant with the phrase "character complex." But anthropologists and human anatomists (perhaps from lack of experience in the practice and principles of taxonomy) often tend to focus their attention on single characters in their discussion of relationships; or, if they take into account a list of several characters, they tend to treat them as an assemblage of separate individual units without recognizing that in combination they constitute a *functional* pattern that must be treated as a whole. It is to emphasize this most important principle that the term "pattern" is put forward as somewhat equivalent to (but actually meaning more than) the term "complex." Undoubtedly, many of the conflicting opinions expressed in the past by comparative anatomists regarding the relationships of the Hominidae and other Primates have been the result of the separate comparison of *individual* characters. In fact, however, it is doubtful whether any single structural detail or measurement by itself can be accepted as providing a clear-cut distinction that would permit a positive identification of a single specimen of a fossil hominoid. For example, the mere presence of a measurable gap (diastema) between the upper canine and lateral incisor teeth is not by itself sufficient to determine that a fragment of a fossil hominoid jaw is identifiable as that of an anthropoid ape rather than a hominid (though a diastema is very rarely absent in the completely erupted dentition of anthropoid apes and is very rarely found in a hominid jaw). But if a well-marked diastema forms one element of a complicated morphological pattern that also includes conical, projecting, and overlapping canine teeth, a laterally compressed lower first premolar tooth of sectorial form, forwardly placed incisors, and so forth, it becomes an important feature in the taxonomic identification of a fossil specimen, for such a pattern is diagnostic of the Pongidae in contradistinction to the Hominidae. Or, to take another example, the mere presence in one specimen of a primitive type of skull of a "mastoid process," considered simply and solely as a bony eminence, does not by itself identify it as hominid rather than pongid, for a mastoid process *of a sort* may occasionally be found in the skulls of mature gorillas (as Schultz [1950a] has noted). But if a mastoid process of typical hominid shape, disposition, and proportions is found *consistently* in a number of skulls in a common collection of fossil hominoids (immature as well as mature), and if it also forms a component part of a total morphological pattern of associated parts normally found in hominid, but never in pongid, skulls (including a clear-cut digastric fossa and a well-marked occipital groove, and a characteristic relationship to the nuchal area of the occipital bone and to a tympanic plate of a particular conformation), its significance for taxonomic purposes evidently becomes enormously enhanced. Many similar examples could be adduced (and some will later be noted) of a coincidence of total morphological pattern that has been overlooked or obscured by treating in isolation only one or two features of the pattern and assuming that these are by themselves adequate for the assessment of taxonomic affinities.

Parallel and Convergent Evolution

It is a fundamental principle of taxonomy that a closeness of resemblance in total morphological pattern indicates a corresponding closeness in zoological relationship. On the other hand, it is also generally recognized that structural resemblances of a sort can be produced by parallel and convergent evolution, and in this case they are not of equal value in the assessment of relationships. Taxonomists are well aware of these complications and of the need to take account of them. But the potentialities of convergence and parallelism have been much overestimated by some comparative anatomists, who have sought to discount the structural resemblances between *H. sapiens* and the anthropoid apes by attributing most (if not all) of them to these processes. That some degree of parallel evolution has occurred in the Hominidae and Pongidae is not in doubt; but to attribute all those similarities that form component elements of a highly complicated total morphological pattern to long-standing convergence or parallelism reduces the morphological principles underlying taxonomy to an absurdity. By the skillful manipulation of such extreme misapplications of well-known evolutionary processes (and, incidentally, taking no account of the statistical improbabilities at once made evident by a consideration of the principles of genetics), it is possible for anyone to draw those conclusions regarding systematic affinities that conform most to his personal predilections. There is no need here to enlarge on this matter of convergence and parallelism in relation to taxonomy (or on the misconceptions some authorities have held on the degree to which evolution is irreversible), for such questions have recently been dealt with in a thoroughly reasonable manner by paleontologists well qualified to do so by long experience and recognized distinction. But it is worthwhile quoting Simpson (1950*b*) that "the basis of parallelism is initial similarity of structure and adaptive type, with subsequent recurrent homologous mutation"; that the initial similarity and the homology of mutations themselves imply phylogenetic relationship; that "closeness of parallelism tends to be proportional to closeness of affinity"; and that "it is improbable that convergence ever produces literal identity in structure and certainly no such case has ever been demonstrated." The extensive data of comparative anatomy and vertebrate paleontology now available justify the assumption (at any rate as a reasonably secure working hypothesis) that species and genera that show a preponderance of structural resemblances are genetically related forms. This would be falsified only if some flagrant discrepancy were seen to exist in one or more features such as could be explained morphologically only by a long period of independent evolution from an ancestral form of a much more primitive type. Evidence of a recent divergence in certain characteristics would not falsify our assumptions. It may not always be possible to exclude finally the factor of convergence as an explanation of a similarity in individual structural features; but it is not permissible to dismiss a complicated pattern of morphological resemblances as merely the expression of convergence without presenting evidence in support of such an arbitrary explanation. As Colbert (1949) has pointed out, "parallelism should not be invoked to explain resemblances among related animals unless it can be proved, for to do this is to make the whole concept of evolution largely meaningless."

The Multiplication of Genera and Species

Defining the species category in paleontology is a difficult problem. It is essential to begin this discussion by recognizing the existence of the one taxonomic category that describes an easily recognizable biological reality, the "biological species" or *biospecies*. This term, which describes a species of living animals, is defined by Mayr (1970) as referring to "a group of interbreeding natural populations that are reproductively isolated from other such groups." Thus a biospecies (under natural conditions) is a genetic entity, genetically isolated from other species. Adding the dimension of time to this concept transforms the biospecies into a discrete lineage with a time dimension, commonly called a *paleospecies* or *chronospecies*, which arises from, and is continuous with, another ancestral paleospecies. Clearly, the arbitrary factors in the definition of the paleospecies, for paleontologists, are the points in time at which it is considered to begin and end. This is one question for which a consideration of variability is important. The paleospecies will normally be given, by convention, the same amount of variability in its fourth (temporal) dimension as a biospecies is given in the present, at a single point in time (Simpson 1961) (see fig. 2). Though this rule does give us some guide about the appropriate amount of

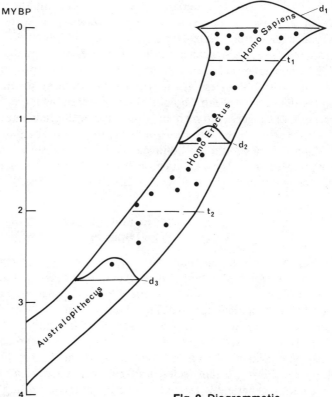

Fig. 2. Diagrammatic representation of successive chrono- or paleospecies of Hominidae during the past four million years (omitting *Australopithecus boisei*): d_1, d_2, d_3, are distribution curves of *H. sapiens*, *H. erectus*, and *Australopithecus* biospecies at certain points in time; t_1, t_2 are conventional time lines drawn to demarcate successive taxa. Dots indicate imaginary fossil specimens (see also Simpson 1961).

variability, it leaves much room for discussion and disagreement, because living biospecies vary in the amount of variability they carry. Argument arising from this problem is a common feature of paleontological discussion, especially among the Hominidae, which have such a full fossil record. In this book an attempt has been made to divide the hominid lineage in a manner which finds wide acceptance among scholars, but which is bound to attract criticism from some workers.

Another vexing taxonomic problem of Primate paleontology is the somewhat arbitrary multiplication of genera and species on the basis of skeletal remains that in many cases are very fragmentary. This, of course, is also a problem of vertebrate paleontology in general. Skeletal elements by themselves do not always reflect to the same degree those differences that are apparent enough in the living animal to warrant distinction at a specific, or even a generic, level. For example, in the Recent Cercopithecoidea the comparative odontologist might not always find it easy to justify—on the basis of the dentition alone—some of the specific distinctions that are clearly justified by a study of the animals as a whole. On the other hand, these very examples may lead him to attribute an exaggerated importance to trivial details of the dentition in *other* groups of Primates in which they may be no more than an expression of individual (or subspecific) variation. Particularly is this the case with the study of fossil remains of the Hominoidea; for the individual, sexual, and subspecific variations in this group of Primates are rather considerable (see Remane 1922). There is thus an almost inevitable (and perhaps, therefore, excusable) tendency among Primate paleontologists too readily to multiply species or genera among the fossil remains they study, if these are only few and fragmentary, for the obvious reason that the extent of the individual and group variability can hardly be assessed until a much more complete record becomes available. But this initial tendency toward "splitting," provided it is recognized, is perhaps not of great importance; and it is often a matter of convenience to accept provisionally and temporarily the generic and specific differentiations that have been made by a paleontologist in the first instance, even though there may be some doubt about their taxonomic validity. Later on, as more fossil material becomes available or as the result of a more detailed comparative analysis of the relevant structural details, the initial taxonomic distinctions may need to be revised. The genus *Sinanthropus* ("Pekin man") was at first based on a single tooth, and the name was retained and employed for many years, even by those who felt dubious about its validity. With further discoveries in China and Java, it became clear that *Sinanthropus* was really not distinguishable generically from *Pithecanthropus* ("Java man"), and it was included in this genus of early hominids. Now, however, even the generic term *Pithecanthropus* has been discarded, for most anthropologists have agreed that the type to which it was applied comes more properly within the genus *Homo*, with the specific distinction of *Homo erectus* (see p. 103). The Chinese representative of this extinct species may still be distinguished at the subspecific level from the Javanese representative, and it is a distinction that is convenient to retain, pending the accession of further material that may allow a more complete study of the limits of variance in the two groups. Another example illustrating the difficulties of taxonomy in the study of fossil remains is furnished by the discovery of *Australopithecus* in South Africa (see p. 129). The first skull to be found was described by Dart (1925), who created for it the specific name *Australopithecus africanus*. Subsequently, Broom found many more remains of individuals belonging to the same group and thought that he was able to distinguish other genera and species—*Plesianthropus transvaalensis* (found at Sterkfontein), *Paranthropus robustus* (found at Kromdraai), *Paranthropus crassidens*, and *Telanthropus capensis* (both found at Swartkrans). Some years later, a new generic term, *Zinjanthropus*, was conferred on an australopithecine fossil found at Olduvai in Tanganyika by Leakey. It is probably true to say that most authorities have now agreed that a convincing case has not yet been made out for separating these fossils generically (see p. 135). However, it may be in some cases a matter of convenience in preliminary discussions to refer to provisional taxonomic terms of this kind, though of course this does not commit those who do so to their final acceptance. On the other hand, where there is a large element of doubt, the wisest procedure is to use place names referring to the site of discovery and to speak (for example) of the "Sterkfontein skull," of the "Kromdraai mandible," or of the "Swartkrans pelvis." In view of the conflicting opinions still held on the nomenclature of the fossil representatives of *Australopithecus*, they are commonly referred to collectively as the subfamily Australopithecinae or, colloquially, the australopithecines.

Variability and Continuity

It is a matter of great difficulty to formulate precisely the criteria by reference to which specific or generic distinctions are recognized in fossil material. Variability due to individual differences, sex, and age will be present in all populations and samples, and across a geographical or temporal range we shall find variability due to local (racial) adaptations or evolutionary change. When the fossil material is scanty, it is not possible to determine the extent of such variability. In such cases the paleontologist can only proceed on the assumption that these limits are of much the same order as in closely related groups where they are reasonably well known. For this purpose a sound knowledge of such related groups (as well as a considerable taxonomic experience) is obviously a prerequisite, for different skeletal and dental characters show different degrees of variation in different groups. Thus even considerable variation in the size of the molar teeth and jaws or in the cranial capacity in a group of fossil hominids should not delude the paleontologist into making specific distinctions on the basis of such characters alone, for the latter are known to show a very wide range of variation in the single species *H. sapiens*. (For example, cranial capacity varies from under 1,000 cc to over 2,000 cc.) On the other hand, similar degrees of variation found among the fossil remains of certain lower Primates may justify a specific distinction insofar as the same characters are known to be less variable in these types. On the basis of such analogies, it is legitimate, though not very desirable, to make specific or generic distinctions as a provisional taxonomic device (even when the fossil material is scanty); but the validity of these distinctions can of course be finally determined only when sufficiently abundant material is available for comparative study among reasonably large samples.

The reaction in recent years against the tendency to multiply genera and species on the basis of insignificant differences in very fragmentary specimens is particularly welcome. For example, since it is now agreed that the morphological differences between *"Pithecanthropus"* and modern man are not adequate to justify a generic distinction, and that the former type should be taxonomically denoted as *Homo erectus*, there can no longer be any justification on morphological grounds for separating *"Paranthropus,"* *"Zinjanthropus,"* and other australopithecines from the genus *Australopithecus*. They may, of course, represent different species of this genus, but in some cases even this degree of distinction seems doubtful.

In an illuminating essay, Simons (1963) has referred to the claims occasionally made by those who have discovered fossil relics of early man in one region that it is in that particular locality that the evolutionary origin of man took place. The author first emphasizes that the oversplitting of fossil types of hominids and pongids by taxonomists has in large part been due to the tendency to regard geographical separation, and even moderate differences in geological age, as indicative of genetic incompatibility in spite of close morphological resemblances. But, as he points out, during the geological periods of the Miocene and Pliocene—that is, at a time when the Hominidae and Pongidae were undergoing a progressive diversification from a common ancestral stock—there were considerable faunal migrations to and fro between Asia, Europe, and Africa. Among other mammalian types these faunal interchanges included extinct anthropoid apes as well as possible forerunners of man, and they involved not only single genera but in some cases probably single species. It thus no longer makes sense to postulate that "man" evolved in one particular and restricted geographical locality—such as in India or the Far East, or in Europe, or in

North, East, or South Africa. The progenitors of the Hominidae and the earliest representatives of this family may well have been distributed as widespread interbreeding communities over several of these regions.

It seems sometimes to have been assumed (in referring to fossil specimens) that specific or generic distinctions are disallowed if intergradations between these taxonomic categories are demonstrable. However, such a criterion is clearly not applicable to a temporal sequence (nor, indeed, is it always so in a contemporary series of geographically or ecologically segregated groups). If a complete fossil record of the Hominidae were available, there would presumably be complete intergradations from the earliest to the latest representatives of this evolutionary sequence. This would not eliminate the need to make taxonomic distinctions in order to give expression to the successive evolutionary phases in a single sequence. Thus the evidence available at present suggests that some populations of the genus *Australopithecus* gave rise to the genus *Homo*. This implies that at a single point in time we have to make a rather arbitrary boundary where the name of the sequence changes from *Australopithecus* to *Homo* (Campbell 1973), although the process of evolutionary change is continuous and gradual. We also need to separate taxonomically the various subsidiary radiations of the sequence, though naturally the distinctions become less and less obvious as the paleontological record leads back to the *initial* stages of evolutionary segregation of each radiation. We can predict that at the initial phase of such a divergent radiation, the fossils will not be easily segregated and may lead to taxonomic controversy.

Intergeneric gradations in a temporal sequence are well exemplified in the Equidae, for Simpson (1951) remarks, in reference to *Hipparion* and related forms, that the difference from the ancestral type *Merychippus* "is clear-cut and, indeed, obvious when the most characteristic forms are compared, but the change was gradual and even an expert is puzzled as to where to draw the line in the continuous series from advanced *Merychippus* to primitive *Hipparion, Neohipparion*, or *Nannippus*." On the other hand, the discovery of intergradations in the fossil record is of

importance insofar as they facilitate (or confirm) natural groupings in the major taxonomic categories, such as families, sub-
orders, and orders, for they may betray systematic affinities that
are not always fully apparent when only the terminal products of
phylogenetic lines are considered. For example, in the classification of the Primates a comparison of the living tarsier with the
existing lemurs suggests a contrast so pronounced as to justify a
subordinal distinction between tarsioids and lemuroids. But a
study of prosimians of Eocene date has demonstrated that, in
their dentition and skull, these early Primates may present such a
mixture of characters as to suggest that tarsioids and lemuroids
have originated (independently of other groups of Primates) from
a common ancestral stock. This inference finds expression in the
inclusion in Simons's classification of tarsioids and lemuroids in
the same suborder, Prosimii; but their early evolutionary segregation is at the same time reflected in their taxonomic separation
into different infraorders. So far as the Hominoidea are concerned, the taxonomic association of the Hominidae and the
Pongidae in a common superfamily, Hominoidea, depends partly
(as already noted) on the discovery of fossil remains showing what
appears to be a considerable degree of structural intergradation
between the two families.

The Quantitative Assessment
of Taxonomic Relationships

In order to assess general taxonomic relationships by estimating
morphological resemblances accurately, attempts have from time
to time been made to estimate degrees of resemblance (and thus,
it is assumed, degrees of affinity) on a quantitative basis. This
biometrical approach is an attempt to facilitate and place on a
strictly objective basis the comparison of one fossil sample with
another. But unfortunately it is fraught with the greatest difficulties, the main one of which, no doubt, is the impossibility by
known methods of weighting each individual character according
to its taxonomic relevance. If the measurements of every single
morphological character of skull, dentition, and limb bones were

of equal value for the assessment of zoological affinities, it might be practicable to assess the latter in strictly quantitative terms. But it is very well recognized that this is by no means the case. It is well known also that the products of parallel evolution may lead to similarities (particularly in general overall measurements and indices derived therefrom) that, if expressed quantitatively, would give an entirely false idea of systematic proximity. Generally speaking, it is true to say that statistical comparisons of overall measurements and indices are of the greatest value in assessing degrees of affinity in forms already known to be quite closely related—for example, subspecies or geographical races—but they become of less and less practical value as the relationship becomes more remote and the types to be compared become more disparate. This was made clearly evident in the pioneer studies of Pearson and Bell (1919) on the femur of the Primates. For though their comparisons of the dimensions and indices of the femur in Recent man (*H. sapiens*) and his Paleolithic precursors (e.g., Neandertal man and *H. erectus*) clearly provide data of considerable value for assessing their relative affinities, the phylogenetic interpretations they give to their comparisons with lower Primates not only are very meager and tentative but are not always in accord with the paleontological evidence that has since become available.

Since the original studies of Pearson and his colleagues, the application of biometrics to taxonomic inquiries has become commonplace. But, because statistical methods are sometimes applied uncritically and without due appreciation of the morphological and phylogenetic basis of taxonomy or of the fundamentals of the phenomena underlying the data to be measured, they have been open to criticism and are in serious danger of becoming discredited. For this reason it seems worthwhile to draw attention to a number of fallacies that are often overlooked by workers in this field, particularly by anatomists and anthropologists who are not morphologists by training or who are not statistical experts. They are listed here in the earnest hope that, by their careful avoidance in the future, some common sources of confusion may be avoided in discussions on the evolutionary origin of man.

The Fallacy of Relying on
Inadequate Statistical Data

One of the limitations of the biometric analysis of taxonomic characters depends on the fact that, if adequate statistical methods are employed, the analysis of even a few measurements entails a very considerable amount of work. Consequently, there is a danger of relying on too few measurements, a danger that is of course seriously increased if these happen to have little taxonomic relevance. The comparison of such measurements may lead to the statement that, say, a fossil bone or tooth shows no significant difference from that of *H. sapiens,* or perhaps from that of the Recent anthropoid apes. But, clearly, such a statement is of doubtful value (and may actually be very misleading) if at the same time account is not taken of other morphological features that may in fact be much more relevant for assessing affinities. An example of this difficulty is provided by the famous case of *Hesperopithecus.* This generic name was given to a fossil tooth found in Nebraska in 1922, on the assumption that it represented an extinct type of anthropoid ape. Part of the evidence for this assumption was based on a comparison of the overall measurements of the tooth with a series of ape teeth, for these metrical data established clearly that in this respect the fossil tooth falls within the range of variation shown in Recent apes (Gregory and Hellman 1923). However, it was the critical eye of a comparative anatomist, with a long experience of the examination and discrimination of paleontological material, that drew attention to certain "nonmetrical" morphological details throwing serious doubt on the original interpretation. As is well known, the tooth later proved to be that of a fossil peccary. This example, which has a certain historical interest in the field of Primate paleontology, is quoted here not in criticism of those who were responsible for the mistaken identification, but to emphasize that two or three overall measurements of a tooth can express only an insignificant proportion of all those metrical elements that contribute to its shape as a whole. This applies also, of course, to skulls or individual bones.

*The Fallacy of Treating All
Metrical Data as of Equal
Taxonomic Value*

It has already been emphasized that morphological characters vary greatly in their significance for the assessment of affinities. Consequently, it is of the utmost importance that, in applying statistical methods, particular attention be given to those characters whose taxonomic relevance has been duly established by comparative anatomical and paleontological studies. This *principle of taxonomic relevance* in the selection of characters for biometrical comparisons is of great importance, but it is also rather liable to be overlooked. One may ask how the distinction is to be made between morphological characters that are relevant or irrelevant for taxonomic purposes. The answer to this question is that each natural group of animals is defined (on the basis of data mainly derived from comparative anatomy and paleontology) by a certain pattern of morphological characters that its members possess in common and that have been found by the pragmatic test of experience to be sufficiently distinctive and consistent to distinguish its members from those of related groups. The possession of this common morphological pattern is taken to indicate a community of origin (in the evolutionary sense) of all the members of the group, an assumption whose justification is to be found in the history of paleontological discovery. But, as a sort of fluctuating background to the common morphological pattern, there may be a number of characters, sometimes obviously adaptive, that not only vary widely within the group but overlap with similar variations in other groups. Such fluctuating characters may be important for distinguishing, say, one species from another within the limits of the family, but they may be of no value by themselves for distinguishing this family from related families. In other words, they are taxonomically irrelevant so far as interfamilial relationships are concerned. The same applies to other major taxonomic categories such as superfamilies, subfamilies, and so forth. For example, among the lemurs the overall dimensions (length and breadth) of the molar teeth may provide useful criteria for distinguishing between the various species and subspecies of the Galaginae or between those of the Lorisinae, but they could not be expected to be of any value in differentiating between these two subfamilies.

So far as the Hominidae are concerned, the principle of taxonomic relevance may be illustrated by reference to the extinct species *Homo erectus* (see p. 118). The available evidence indicates that in this type the morphological features of the skull and jaws are very different from those of *H. sapiens*, whereas the limb skeleton is hardly distinguishable. Clearly, therefore, if the question arises whether the remains of a fossil hominid are those of *H. erectus* or *H. sapiens*, for taxonomic purposes the morphological features of the skull and jaws are the relevant characters to which attention should be primarily directed. In the study of fossils representing early phases in evolutionary radiations, their affinities must be determined by a study of those characters whose taxonomic relevance may be inferred from a consideration of the main trends of evolution as demonstrated by comparative anatomical studies and by extrapolation from the fossil record, so far as the latter is available. For example, as we have already seen, the initial evolutionary segregation of the Hominidae from the Pongidae was almost certainly dependent on modifications related to the development of an erect bipedal gait (Washburn 1950). Hence, in assessing the affinities of the *earlier* representatives of the Hominidae (whose taxonomic position may be in some doubt), the skeletal characters of the pelvis and hind limb are likely to be of much greater importance than those of the forelimb. As we shall see later, also, the morphological details of the dentition are likely to be of much greater taxonomic relevance than the actual overall dimensions of the teeth and jaws or the cranial capacity. As a further example of the principle of taxonomic relevance, we may refer to the dentition of some of the fossil representatives of the Pongidae. In these the incisor teeth are so similar to those of *Homo* (and even *H. sapiens*) as to be hardly distinguishable. On the other hand, in all known pongids the canine teeth are quite different. Obviously, therefore, in distinguishing between the pongid or hominid affinities of a fossil hominoid, the canines have a much higher degree of taxonomic relevance than the incisors.

It perhaps needs to be emphasized that the principle of taxonomic relevance must also, of course, be taken into account in any attempt to assess the affinities of a fossil type from the biometrical study of a single skeletal element. For not all the dimensions or indices of such a specimen will have the same taxonomic relevance, and some may have none at all for the particular comparison under consideration. The pelvic bone of the fossil Australopithecinae from South Africa (see p. 163) provides a good example of this point, and it also illustrates the essential importance of distinguishing between those morphological characters that may be similar in two divergent evolutionary groups simply because they are inherited from a common ancestry and those characters that represent adaptive modifications peculiar to, and are thus distinctive and diagnostic of, one or the other of the two groups. The former type of character is obviously not taxonomically relevant for distinguishing the two groups (at least in the earlier stages of their evolutionary development); the latter type evidently has a high degree of taxonomic relevance. The australopithecine pelvic bone presents a most interesting combination of characters. Some of these are quite distinctive of the hominid (as opposed to the pongid) line of evolution, such as the width-height ratio of the ilium, the development of a strong anterior inferior iliac spine, the orientation of the sacral articulation, the formation of a deep sciatic notch, and so forth; and together they compose a morphological pattern that is evidently an adaptation to the mechanical requirements of an erect posture. On the other hand, there appear to be no characters that are definitely distinctive of the pongid (as opposed to the hominid) line of evolution. It is true that in a few features the pelvic bone is distinctive from that of modern man, and in these particular features it does show some degree of resemblance to that of the modern anthropoid apes. But such a resemblance is clearly due to the retention of primitive features derived from a common hominoid ancestry and is thus not indicative of any close affinity with the modern anthropoid apes; this is made quite clear by the fact (as we have just noted) that, in those features in which the australopithecine pelvis *has* undergone modification away from the primitive ancestral type, the modification has followed the direction of hominid evolution and not of pongid evolution. It is these latter (positive) features, therefore, that are relevant for determining the taxonomic status of the Australopithecinae so far as the pelvic bone is concerned.

To keep within reasonable limits the number of measurements to be used for the statistical comparison of a fossil bone or tooth with related types, the rational procedure is first to make direct visual observations, selecting for comparison just those features that are known to have taxonomic value for the problem in hand. In many cases differences or resemblances may be so obtrusive as to obviate the need for statistical methods altogether. On the other hand, if differences and resemblances are not immediately apparent on visual inspection, special ad hoc measurements and indices may then be devised in order to test those characters that can reasonably be expected to be of value in the assessment of systematic affinities in any particular case. Only negative results are to be anticipated if routine measurements of little or no taxonomic value are employed.

The Fallacy of Treating
Characters Separately and
Independently instead of in
Combination

This fallacy has been treated in some detail by Bronowski and Long (1952). They point out that a bone or a tooth is a unit and not a discrete assembly of independent measurements, and that to consider their measurements singly is likely to be both inconclusive and misleading. The right statistical method, they emphasize, must treat the set of variables as a single coherent matrix. This can be done by the technique of multivariate analysis, which is essentially a method (not possible with more elementary techniques) that can be used for comparing morphological *patterns*. In principle, the application of the technique is straightforward enough, but it requires care and discrimination, a sound knowledge of morphology, and also a considerable experience of statistical methods. A number of measurements or indices of a bone or tooth are selected, which are judged on morphological grounds to be taxonomically significant; and from these the averages, variances, and correlations for a number of specimens are calculated. It is then possible to construct a numerical picture of the size and shape of the bone or tooth (and of the extent to which they vary) and to express this as a discriminant function. Such functions may be used for deciding whether, say, a fossil hominoid tooth is more likely to belong to a pongid or a hominid type, provided of course that the particular discriminant functions already calculated for the two families are sufficiently distinct. Bronowski and Long emphasized the value of multivariate analysis by applying it to a controversial issue that had arisen in regard to certain teeth of *Australopithecus* and they were able to resolve the controversy by demonstrating very positively their hominid character.

The Principle of Morphological Equivalence in Making Statistical Comparisons

Failure to understand this principle is perhaps one of the most serious sources of fallacy likely to affect statistical studies by those who are not thoroughly acquainted with the morphology of the skeletal elements they are dealing with. A simple (but rather crude) example may be offered by referring to a measurement often employed in craniology—the auricular height. This is commonly taken by measuring the maximum height of the skull (in the Frankfurt plane)[2] from the auditory aperture; and in comparing different racial groups of *H. sapiens* it gives an index of the height of the braincase at this particular level. But, in comparing *H. sapiens* with, say, the gorilla, it would clearly be misleading to employ the same technique, for in male gorillas the height of the skull is often considerably extended by the development of a powerful sagittal crest. If such a comparison were made, it would be a comparison of the height of the braincase in *H. sapiens* with the height of the braincase *plus* a sagittal crest in the gorilla and would have no meaning from the morphological viewpoint. This is, of course, an extreme example, but it is perhaps not fully realized that similar, if less obvious, fallacies may be incurred in other craniometric work in which overall measurements of the skull are commonly equated with one another. In comparing skulls of closely related groups, such measurements may be sufficiently equivalent morphologically to make direct metrical comparisons valid. But if they are used to compare, say, a modern European skull with the skull of the fossil species *H. erectus*, serious difficulties are involved. For example, in the European the glabellomaximal length is an approximate measurement of the maximum length of the braincase. But in the skull of *H. erectus* it measures a good deal more, for the glabellomaximal length is complicated by the exaggerated development of a massive supraorbital torus, the great thickness of the skull, and the projection backward of an exaggerated occipital torus. The overall maximum length measurement is thus not strictly comparable (in the morphological sense) with that of a European skull—in both cases it involves a number of different elements that may be independently variable among themselves. Again, in the European skull the maximum width is commonly situated in the parietal region, whereas in the *Homo erectus* skull it is situated in the temporal region (see fig. 22, p. 154). Thus to compare the maximum width in the two skulls is to compare measurements that are also not morphologically equivalent. In fairness to physical anthropologists generally, it must be stated that these sources of fallacy in comparative osteometric studies are usually very well recognized, but this may not always be the case. As a further example we may take the lower front premolar tooth (P_3) in the Hominidae and Pongidae. One method that has been used for measuring the length of this tooth is to take the maximum anteroposterior diameter in the axis of the tooth row.

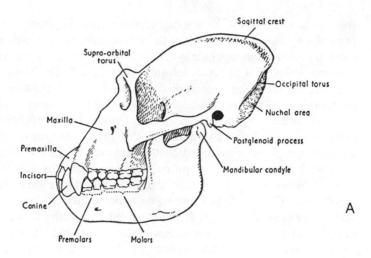

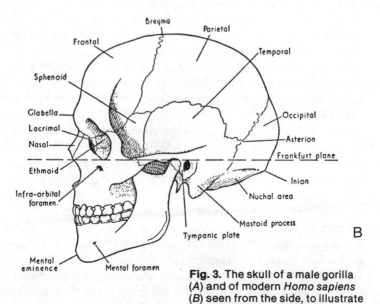

Fig. 3. The skull of a male gorilla (*A*) and of modern *Homo sapiens* (*B*) seen from the side, to illustrate some of the anatomical landmarks referred to in the text.

But, as is well known, in the anthropoid apes the front lower premolar is commonly rotated on its vertical axis, so that the axis corresponding morphologically to the transverse axis of the hominid premolar is directed obliquely posteromedially. Thus, to compare the maximum anteroposterior diameters in the axis of the tooth row in the two groups is to compare dimensions of the premolar which again are not morphologically comparable. What is actually being compared in this case is the maximum anteroposterior *space* occupied by the premolar in the tooth row —a very different thing. In comparing overall measurements of skeletal elements, it is of the greatest importance that the morphological basis of the dimensions to be compared should be stated very precisely indeed. It may be argued that the fallacy of morphological nonequivalence must almost necessarily be involved in any overall measurement of a skeletal structure, since such a measurement is bound to include a number of different components that may vary independently. For example, two skulls may show the same thickness of cranial wall, but the latter may actually be composed of different proportions of the outer table, inner table, and diploe. This, of course, is perfectly true, but it only serves to emphasize still more strongly the need for care in stating the morphological fundamentals of the biometric data employed.

The amateur biometrician, when comparing bones or teeth of different shapes, has sometimes fallen into the trap of comparing dimensions that (because of the different shape) are not morphologically comparable and then, on the basis of this false comparison, concluding that because these dimensions are similar the bones or teeth are actually of the *same* shape. We may reiterate here that, though biometrical studies of immediately related forms belonging to the same restricted group (such as a species or subspecies) may be expected to give fair comparisons that approximate closely enough in their morphological equivalence, the statistical comparison of different genera that show a greater disparity of form needs to be carried out with a very critical appreciation of the technical difficulties involved.[3]

*The Fallacy of Comparing
Skeletal Elements in
Individuals of Different Age,
Sex, and Size*

It is well recognized that age changes may lead to considerable modifications in the structural details and proportions of skeletal elements. In the skull, for example, they are so marked that it would clearly be fallacious to compare a few measurements of the adult skull of a primitive hominid with those of juvenile skulls of an anthropoid ape and to infer from such a comparison that the former is not markedly different from anthropoid apes in general. In regard to sex differences, again, it would obviously be misleading to compare the dimensions of the canine teeth in presumably male fossil hominoids with the relatively small teeth of a female gorilla and to conclude therefrom that, in these particular dimensions, the fossil teeth fall within the range of variation of those of Recent apes. Sexual dimorphism needs to be taken into account in such a case, as it does also in comparisons of morphological and metrical features of the skull and skeleton in general.

The factor of body size in statistical comparisons is perhaps of even greater importance because it has been overlooked much more frequently by physical anthropologists. Differences of proportions in the skull and skeleton in Primates of different sizes may be merely an expression of allometric growth (see, for example, Pilbeam and Gould 1974), or they may be related to the mechanical requirements dependent on differences in body weight. In either case, of course, they may be of very little taxonomic importance (except perhaps in the determination of specific or subspecific distinctions). Thus in quadrupedal mammals the relative thickness of the leg bones is a function of the absolute size of the animal, for the strength of a bone as a supporting structure varies as its cross-sectional area (i.e., as the square of the linear dimensions of the animal), while the weight of the animal varies as the volume (i.e., as the cube of the linear dimensions). In heavier mammals, therefore, the leg bones are relatively thick, and their actual shape may thus be markedly different from those of lightly built (but still closely related) types. Clearly, then, it may be very misleading to compare the "robusticity index" of, say, the femur in Primates of very different body sizes (e.g., hominids, apes, and monkeys) and then assume degrees of affinity or divergence without any reference at all to the body-size factor. The differences in shape of bones may be accentuated even more by the fact that in larger animals the muscular ridges, tuberosities, and so forth are much more powerfully developed. Nor does this difficulty apply only to limb bones. In the skull it is well known that in closely related animals account must be taken of the factor of allometry in comparing relative size of braincase and relative size of jaws; and variations in the proportion and indices of these structures may again be reflected in differences depending on the degree of development of muscular ridges or of bony features that have developed in response to mechanical stresses. Thus, for example, it would be futile to compare cranial indices of a gorilla with those of a small monkey with any idea of drawing taxonomic conclusions unless factors of body size are first taken into account. For the same reason (though at first sight the case is less obvious) it would be misleading to make a direct comparison of cranial indices in the small and delicately built skull of a pygmy chimpanzee with those of a large and massive skull of a fossil *Australopithecus*. And even in comparisons of modern human skulls of a single homogeneous series, account must be taken of absolute size, for it is well established by biometrical studies that there is a significant correlation between form (as expressed in cranial indices) and absolute size.

It can hardly be overemphasized that in comparing dimensions and indices of skull, limb bones, pelvis, or other skeletal elements of Primates generally, taxonomic conclusions must be preceded by an inquiry into all the complicating factors related to body size, an inquiry that may need elaborate statistical studies and that most certainly requires an intimate knowledge of the structural responses of skeletal elements to functional demands.

*The Fallacy of Comparing
Measurements Taken by
Different Observers Using
Different Techniques*

The dangers of this fallacy have been emphasized again and again by biometricians, and the only excuse for mentioning it here is that it is still overlooked by some writers. If it is certain that the different observers are using identical techniques for recording their measurements, the latter may be employed for comparative studies; but they still need to be used with the greatest circumspection (particularly in the case of very small objects such as teeth, where measurements need to be accurate to a tenth of a millimeter). Where it is apparent that different observers are not employing precisely the same techniques, statistical comparisons must necessarily be stultified. In anthropological craniometry attempts have been made (with some success) to secure general agreement on the definitions of points and planes that serve as a basis for statistical measurements (see, for example, Buxton and Morant 1933). In regard to some of the other elements of the skeleton and also the dentition, the virtual absence of standardization of metrical technique renders comparisons between different observers very hazardous indeed. The definition of such measurements should always be carefully investigated.

*The Fallacy of Relying for
Assessment of Affinities on
Biometrical Analysis of
Characters That May Have No
Genetic Basis*

Anthropologists do not always recognize that during the period of growth bone is a very plastic material. That is to say, its form may be readily modified by the mechanical effects of pressure and traction of the soft parts immediately related to it and also by the effects of dietetic deficiencies or constitutional disturbances of one sort or another. This needs to be taken into account as a possible source of fallacy in the attempts that have sometimes been made to assess the affinities of the various racial groups of *H. sapiens* on the basis of osteometric data, particularly in those cases where differences are so slight as to be detected only by statistical methods. So far as paleoanthropology is concerned, also, it is a factor that always needs to be taken into account when seeking evidence for the differentiation of geographical variants of the same general type. For example, it has been argued that the Javanese and Chinese representatives of the Pleistocene species *H. erectus* are taxonomically distinct because (inter alia) the thigh bone of the latter shows a flattening of the shaft (*platymeria*) that is not present in the former. But, apart from the fact that this feature shows considerable variation within the limits of the single species *H. sapiens*, there is some indirect, suggestive evidence that the degree of flattening of the shaft may depend on nutritional factors (Buxton 1938). If this is so, it is not a character that can be properly used for taxonomic reference without qualification. How far nutritional or other postnatal influences may determine minor differences in cranial or facial proportions is still very uncertain. It is for this reason, of course, that in the study of modern populations physical anthropologists are now placing less reliance on the comparisons of traditional anthropometry and are concentrating their attention on biochemical characters whose genetic composition is more directly ascertainable and by reference to which racial groups can be classified objectively on the basis of gene frequencies.

The Importance of Geological Age for Determining the Evolutionary Position of Fossils

It is necessary to recognize the distinction between a morphological series linking one taxonomic group with another and a geological series representing a true temporal sequence. A graded morphological series, by itself, may be of direct importance in determining major taxonomic groupings, and it may also provide indirect evidence of great value for a provisional assessment of phylogenetic relationships. But evolutionary lines of development can be finally determined only by the demonstration of an actual temporal sequence, and the latter can be established only if geological dating is secure. It is necessary to emphasize that anatomical studies *by themselves* can, of course, provide no basis for assessing geological age—this must depend on the studies of geologists based on stratigraphic data and the estimation of radioactive and other chemical elements in fossilized remains, of paleontologists based on faunal evidence, and of archeologists based on cultural sequences. Of these different lines of evidence, that provided by geological data is undoubtedly the most important, for paleontological and archeological evidence of the antiquity of fossilized remains is essentially derivative, being itself ultimately based on geological evidence. So far as hominid paleontology is concerned, it is unfortunate that the geological evidence of antiquity has in the past so often been equivocal, mainly for the reason that the fossil remains were discovered by chance and the stratigraphic evidence obscured before there had been an opportunity for a systematic study of the site, or they were found in the course of excavations by workers who were not qualified by training and experience to assess the geological evidence.

It is unfortunate that in spite of the progress made in recent years in radiometric dating techniques (see pp. 60–61) there is still much uncertainty about the age of many fossil-bearing deposits. In the north temperate regions this problem is particularly acute because most of the finds come from a period and geological context not suitable for existing radiometric methods. As a result the discoveries have in most cases been correlated with the succession of glacial periods that has characterized north temperate Eurasia since the beginning of the Pleistocene. Recent research finally makes it clear that there have been numerous climatic oscillations that have been triggered by variations in the earth's orbit (Hays, Imbrie, and Shackleton 1977), producing a large number of glacial and interglacial periods lasting from 5,000 to perhaps 50,000 years each. The most easily recognized of these have been given names, in Europe and elsewhere, during the past seventy years. But it is not yet possible to correlate these named periods with the various oscillations recognized by Emiliani and dated through the intensive study of deep-sea cores.[4] It seems possible that in the next few years the recognized series of four glacial epochs, most generally known by the names Gunz, Mindel, Riss, and Würm, will be replaced by a numbered and dated sequence of something over twelve glacial phases, but the equivalence of the two systems will not be easily determined. In this book, therefore, it has proved most convenient to stick to the old European four-stage terminology with the caveat that there is still much uncertainty attached to it and that it gravely oversimplifies the glacial succession. Figure 4 (p. 54) shows a very rough approximation of the absolute age of these well-known, but not well-founded, glacial phases and how they might be correlated with Emiliani's oscillations.

Undoubtedly many of the unsatisfying controversies that in the past have been aroused by the discovery of hominid fossils might have been avoided had it been possible in the first instance to establish a date with reasonable assurance. In the absence of any degree of certainty, there has been a tendency for some anatomists to select that evidence for antiquity that seems best to fit in with the morphological status of the fossil. But this tendency must be strenuously avoided, for it introduces a very obvious subjective element. Naturally, if the anatomist finds that a skeleton assigned to a great antiquity shows no significant difference from that of modern *H. sapiens*, he is entitled to issue a caveat or perhaps to demand a more rigorous inquiry into the evidence for geological age, but he is not entitled to ignore the evidence simply because it conflicts with his preconceived ideas of evolutionary history. On the other hand, if a fossil of primitive type is found in a deposit that is *later* in geological age than might be expected on the general evidence already available, this need not disturb conclusions previously accepted regarding its evolutionary position, for it is well recognized that archaic types may persist for long periods in some parts of the world after they have given rise to more advanced types elsewhere (see fig. 1). For example, individuals of the species *H. erectus* certainly persisted in the Middle Pleistocene in the Far East at a time when early *H. sapiens* was already in existence in Europe. Similarly, one or more species of *Australopithecus* certainly persisted long after another had given rise to the genus *Homo*. A parallel example is also to be found in the evolutionary history of the Equidae, for some representatives of the genus *Parahippus* are known to have survived into the Upper Miocene, although other representatives of the *same* genus had already given rise to the succeeding phase of equid evolution (*Merychippus*) by the Middle Miocene.

These considerations have reference to a very common source of confusion and needless argumentation in discussions on evolution (and particularly on hominid evolution) when the suggestion is made that a particular fossil type may be "ancestral" to a later type. Thus, if the proposition is put forward that *H. erectus* is ancestral to *H. sapiens*, this does not, of course, mean (though some critics have evidently supposed it to mean) that the particular individuals whose remains have actually been found in Java or China are claimed to be the direct ancestors of *H. sapiens*, or even that the local group or subspecies of which they are representatives included direct ancestors. It means only to suggest that the species as a whole provided the matrix that gave rise—perhaps elsewhere in the world and perhaps at an earlier time—to the precursors of *H. sapiens*. Or, to put it another way, it means that, so far as their probable morphological characters can be inferred from comparative anatomical and paleontological studies and also by analogy with what is known of the evolutionary history of other mammalian groups, the ancestors of *H. sapiens* would have resembled the known individuals of the *H. erectus* group so closely as not to be specifically distinguishable. Similarly, when it is suggested that *Australopithecus* may be ancestral to *Homo* (and on the purely morphological evidence the suggestion is a perfectly valid one), it is not implied that the known representatives of this genus found in Africa are themselves the actual ancestors. The fact is that in their anatomical structure the African fossils conform so closely to theoretical postulates for an intermediate phase of early hominid evolution (based on indirect evidence) as to lead to the inference that the actual ancestral group could hardly be *generically* distinct. It is here a question of weighing the available evidence and estimating probabilities. Only when the paleontological record becomes more fully documented by further discoveries will it be permissible to make more definite statements on these particular problems. But, in any case, the chances of finding the fossil remains of *actual* ancestors, or even representatives of the local geographical group that provided the actual ancestors, were it possible to identify them, are so remote as not to be worth consideration.

Primitive (or Generalized) and Specialized Characters

In discussions on possible relationships of certain fossil hominids to *H. sapiens*, the argument is sometimes advanced that they are too specialized in one or another anatomical feature to have provided an ancestral basis for modern types. The suggestion here, of course, is that the development of some morphological character that is not present in *H. sapiens*, and wherein the latter appears to preserve a more primitive condition, implies an aberrant specialization that precludes any consideration of an ancestral relationship. In order to discuss the validity of such arguments, it is clear that we need some definition of the terms "primitive or generalized" and "specialized," and we also need to consider whether a structural specialization by itself necessarily implies the impossibility of reversion to a supposedly more primitive condition.

In studies of phylogenesis the distinction between morphological characters that are essentially primitive or generalized and those that may be regarded as divergent or specialized is commonly based on several considerations. In the first place, a study of the comparative anatomy of living types, particularly of those that, on the whole, are the most simply organized and occupy the most lowly position in a scale of increasing elaboration, may give an indirect indication of the relation between primitive and specialized features. A reference to the earlier fossil representatives of the group will provide further evidence that may indeed be conclusive if the fossil record is sufficiently complete to provide a closely graded sequence of evolutionary development. Detailed anatomical studies of the morphological characters concerned, with special reference to their embryological development, may add still further information. As a simple and very obvious example, we may take the morphological character of penta-dactyly. This is judged to be a primitive and generalized condition in mammals that has in some cases been replaced by specializa-tions depending on the loss of one or more digits; the reasons for this assumption are as follows: in living mammals of a simple organization, and also in Reptilia (from which mammals were originally derived), pentadactyly is the general rule; paleontology has demonstrated that in the early precursors of those mammals that today have fewer than five digits pentadactyly was also a characteristic feature; and, last, a detailed anatomical study of mammals with fewer than five digits may reveal, in the adult, vestigial remains of those that have been lost in the course of evolution or, in the embryo, transient traces of these vanished structures.

By the application of such criteria, it is usually possible to determine which characters can be designated as primitive or generalized and which are obvious specializations. For example, so far as the Hominoidea are concerned, it is certain that the gross elongation of the forelimbs in the Recent anthropoid apes (associated with retrogressive changes in the pollex and modifi-cations of the limb musculature) are specializations from the primitive or generalized Primate condition. The same may also be said of the profound modifications of the skull and skeleton for erect bipedalism in the Hominidae. It may also be accepted, by the same line of reasoning, that the brachydont canine or the retrogressive changes in the last molar of *H. sapiens* are special-ized features.

The importance of making a general distinction between primitive and specialized characters depends on the fact that the latter may be taken to indicate divergent trends of evolution, giving rise to more or less aberrant groups, and such aberrant groups, of course, are unlikely to bear an ancestral relationship to later-evolved groups in which similar specializations are absent. This consideration introduces us again to the much-discussed question of the irreversibility of evolution. As already mentioned (p. 18), there is no need to debate this question here, for it has been adequately discussed elsewhere (Simpson 1950*b*). The important point to recognize is that, although the principle of the irreversibility of evolution is perfectly sound in its general application, it is not legitimate to use it as an argument against ancestral relationships in reference to isolated characters that may have quite a simple genetic basis and that are not obviously related to any marked degree of *functional* specialization. It is well known that some mutational processes may be reversible, for this has actually been demonstrated by genetic studies in the laboratory. It is also certain that individual morphological or metrical characters considered as isolated abstractions may undergo an evolutionary reversal; for example, the size of a tooth may increase and subsequently undergo a secondary reduction (as demonstrated in the paleontological record of the Equidae), or muscles may be developed as new morphological units and later disappear when the functional demand for them ceases to exist (as demonstrated by vestigial or atavistic appearances sometimes seen in the human body). Instances of the loss of a character previously acquired are common enough in paleontology and may be termed "negative reversals" of evolution. On the other hand, a "positive reversal," that is to say, the reacquirement of a complicated or composite morphological character in its *exact* original form after it has been lost in the course of evolution, must certainly be a rarity. For if the initial development of such a character has been the result of a multiplicity of mutations or of a prolonged sequence of successive mutations, and if it has also been dependent on a great complexity of selective influences, the chances of its redevelopment will be so entirely remote as to be discounted altogether as a possibility. Reference may also be made to the line of argument followed by Ford (1938), who points out that as any evolving group becomes more and more specialized in adaptation to one particular mode of life the possible variations that could be of use to it become progressively restricted. "Finally," he goes on to say, "it attains a state of 'orthogenesis' in which the only changes open to the species are those which push it along the path it has already pursued." In other words, it becomes more and more difficult, on the basis of the natural selection of heritable variations, for an evolving line to retrace its steps and thus reverse its evolutionary trends. It need hardly be said that, in using the term "orthogenesis," Ford is referring not to the effects of an inherent tendency within the organism to evolve in a certain direction, but to the effects of what has been called "orthoselection."

But, though it is legitimate to exclude from ancestral relationship to modern types any fossil group that provides clear morphological evidence of an aberrant development obviously related to any extreme functional specialization, it is equally important to avoid the assumption that any minor deviation from a supposedly primitive condition must necessarily also be exclusive in the same sense. For example, it has been suggested that the development of a sagittal crest on the skull of some of the Australopithecinae from South and East Africa (see p. 153) would debar these fossil types from consideration as possible ancestors of *Homo*. But there is no evidence that a sagittal crest is a morphological entity with a separate genetic basis; it is no more than the secondary result of a growth process depending on the combination of a small braincase with large jaws and large temporal muscles (it is "built up" by the further extension upward of these muscles when they have reached the limits of the cranial roof at the midline of the skull). Indeed, it is a character that would be expected to be present in the earlier, small-braincd representatives of the Hominidae, for only as a result of the later expansion of the brain (and the concomitant reduction of the jaws), would the temporal muscles during growth find adequate accommodation on the braincase itself without the need to build up a sagittal crest.

It has also been argued that the Pleistocene hominid species *H. erectus* could not be ancestral to *H. sapiens* because it was characterized by prominent supraorbital ridges that are absent in *H. sapiens* and were presumably absent in the evolutionary precursors of the Hominoidea. But it is too hazardous to draw such conclusions on such slender evidence; for again there is no theoretical or practical reason why, in the hominid sequence of evolution, the development and subsequent disappearance of prominent supraorbital ridges should not be correlated with changing proportions of the jaws and braincase. And, in any case, nothing is known of the genetic basis of this particular morphological character.

Granted that single mutations are reversible in direction and that negative reversals are common phenomena of evolution, the question arises, How are we to determine from the study of the fossilized remains of a group of hominoids whether it has already attained such a degree of specialization in its an. ᵗomical structure that it must be regarded as a divergent or aberrant group having no ancestral relationship to modern types? Surely here it is a matter of assessing the total morphological pattern in terms of the probable complexity of its genetic constitution and of gauging the degree to which morphological changes may have committed the group to a mode of life that has restricted too far the opportunities for selection in other evolutionary directions. Thus, for example, the structural adaptations of the modern anthropoid apes for a brachiating mode of arboreal life—as shown in the modifications of the limb skeleton and musculature—have evidently become too extreme and too complex to permit them (with any probability) to revert to the more generalized structure that would be a necessary prelude to modifications in the opposite direction of erect bipedalism. On the other hand, the large size of the molar teeth of the Australopithecinae and the heavy supraorbital ridges of *H. erectus* could certainly not be regarded as functional specializations of this type. Incidentally, the functional aspect of specialization is perhaps more important for taxonomy than is often realized; for in the assessment of phylogenetic relationships taxonomists have sometimes tended to lay more stress on what are presumed to be nonadaptive characters than on obviously adaptive characters. But, in the first place, it has been questioned whether a character ever is nonadaptive in the sense that it has no relation whatever to the functional demands of the environment. It is true that *by itself* a character may have no selective advantage, but it may be linked genetically with some other character that does. Second, a character that has no selective advantage may much more easily undergo a rapid change by mutational variations than one that is directly adapted to a special environment, for any sudden disturbance of the second type of character might be presumed to place the animal at an immediate disadvantage. In other words, it might be expected that in a given environment obviously adaptive characters would actually be more stable and therefore of more importance for assessing affinities.

We may formulate the general proposition, then, that if a fossil group shows structural changes that are evidently related to (and responsible for) functional specializations, and if the latter appear to have definitely committed it to one special mode of life, it is in the highest degree unlikely that the group would be capable of a true evolutionary reversal. That is not to say, of course, that a specialized group of animals is incapable of changing its mode of life. For example, in the course of evolution an arboreal group may become adapted for terrestrial life and subsequently become adapted again for arboreal life. But in doing so it does not revert to the more primitive or generalized structure of its arboreal ancestors. Its terrestrial specializations are preserved and still further modified away from the primitive condition to allow such a functional transformation. Apart from functional considerations, however, it is also legitimate to discount the possibility of an evolutionary reversal if the morphological divergence from a more primitive condition can be assumed, on the basis of palcontological evidence, to be the cumulative effect of a long succession of small mutational variations exposed to selective influences of a complex nature, or if there is reason to suppose (by analogy from genetic studies of living forms) that the morphological character concerned is the phenotypical expression of a genotype that is so complex that an evolutionary reversal could occur only as the result of a multiplicity of mutational reversals.

In spite of what has been said, it should be emphasized that the ultimate decision whether a fossil genus is ancestral to a living genus or not must be determined by the paleontological record, and this can be done only when the latter is know in sufficient detail. On purely morphological grounds (and without reference to the paleontological sequence), there is no certain argument why the somewhat aberrant West European subspecies of Neandertal man could not be ancestral to *H. sapiens*. But in this particular instance (see p. 65) the fossil record shows with reasonable probability that such was not the case. Some of the early Miocene genera of the Pongidae may have been ancestral to the Hominidae—there likewise appears to be no valid morphological argument against this. But the solution of this particular problem must depend on the amplification of the fossil record by further discoveries.

General Considerations

In this chapter we have been concerned to draw attention to some of the more obvious sources of confusion that all too commonly lead to misunderstandings and misrepresentations in discussions on hominid evolution. Undoubtedly, the failure to recognize the phylogenetic implications of taxonomic terminology has been responsible for much of this confusion, for the reason that there has been a tendency to use this terminology as though it were based on morphological definitions applied to living forms only. The loose employment of colloquial group terms is an even greater source of confusion. It is remarkable that even in strictly scientific papers authors frequently use the terms "man" and "human" without any attempt to define them, and it is clear also that they are used with very different meanings not only by different writers but also by the same writer in different contexts. The term "hominid," again, seems to be used frequently as though it referred only to the large-brained genus *Homo*, whereas it should be strictly employed only as the adjectival form of the taxonomic term "Hominidae." In other words, as we have tried to emphasize, it should be equated with the other familial categories of mammalian classification and apply to the whole sequence of evolution that led to the development of *Homo* (and other hominid genera) from the time when this sequence became segregated from the related family Pongidae.

Assessing genetic affinity by comparing morphological details of no more than portions of a fossilized skeleton (particularly if these are very fragmentary) obviously poses a most serious problem, for in many cases such comparative studies can lead only to inconclusive results. Yet, because of its rarity, it is important that all fossil Primate material be subjected to intensive study and that some attempt be made to assess the taxonomic status of each specimen, even though this may lead to no more than a provisional interpretation of its affinities. In the absence of an abundant fossil record, conclusions regarding lines of phylogenetic development must always be provisional; and, as the evidence accrues with new discoveries, they will need constant revision. Paleontologists themselves are quite aware of this, but it would be well that those less experienced in the study of fossils should recognize it also. It is perfectly justifiable and indeed necessary to put forward working hypotheses on the basis of known evidence: they form the basis of research projects. In due course, new evidence usually requires the modification, if not the abandonment, of such hypotheses. This, however, is how the scientific method works. It amounts to a step-by-step and ever closer approximation to the truth, and implies that all hypotheses, unless they take a very general form (in which case they are of less heuristic value) are necessarily going to be superseded as science progresses.

In summary, it can hardly be emphasized too strongly that, in assessing the taxonomic position of a fossil specimen, account must be taken of the total morphological pattern (and not its individual units) that provides the reliable morphological evidence on which zoological relationships can be determined. Comparing individual characters independently as isolated abstractions, instead of treating them as integrated components of a complex pattern, is perhaps one of the main reasons a multiplicity of systems of classification of the Primates are still to be found in the literature.

Two

Homo sapiens

It is now generally agreed that all the modern races of mankind are variants of one species, *Homo sapiens*. So far as skeletal characters are concerned (and these, of course, are the only anatomical characters available to the paleontologist), this biospecies may be provisionally described as follows:

> *Homo sapiens*—a species of the genus *Homo* characterized by a mean cranial capacity of about 1,350 cc; muscular ridges on the cranium not strongly marked; a rounded and approximately vertical forehead; supraorbital ridges absent or slightly developed and in any case not forming a continuous and uninterrupted torus; rounded occipital region with a nuchal area of relatively small extent; foramen magnum facing directly downward; the consistent presence of a prominent mastoid process of pyramidal shape (in juveniles as well as adults), associated with a well-marked digastric fossa and occipital groove; maximum width of the calvaria usually in the parietal region and axis of glabellomaximal length well above the level of the external occipital protuberance; marked flexion of the sphenoidal angle, with a mean value of about 110°; jaws and teeth of relatively small size, with retrogressive features in the last molars; maxilla having a concave facial surface, including a

canine fossa; distinct mental eminence; eruption of permanent canine commonly preceding that of the second molar; spines of cervical vertebrae (with the exception of the seventh) usually rudimentary; appendicular skeleton adapted for a fully upright posture and gait; limb bones relatively slender and straight.

On the basis of the total morphological pattern composed by these skeletal characters (and also on the basis of biometric comparisons of overall dimensions and of indices constructed therefrom), it is possible to affirm that the populations of late Paleolithic times, so far as they are known from fossilized remains, all conform entirely with those of modern man. In other words, the species as we know it extends back in time at least as far as the beginning of the Aurignacian phase of Paleolithic culture in Europe (see fig. 4),[1] and this, at a conservative estimate, means an antiquity of some 30,000 years. There is no known anatomical feature whereby the skeletal remains of either Europeans, Asians, or Africans of this age can be distinguished from modern man, and, as is well known, the populations in Europe at this early time had already developed a highly complex culture.

Apart from the problem of the antiquity of *H. sapiens* as a species, to which we shall return, is the problem of the antiquity of the modern races of mankind as distinct geographical varieties. The difficult definition of the word "race" and the loose employment of the term in common parlance, as well as the insinuations that have been linked with it by exponents of extreme political creeds, has not made it easy to secure agreement on this matter. It has to be recognized, moreover, that in zoological taxonomy generally there are still problems relating to the definition of the term "species," let alone the definition of infraspecific categories, such as "demes," "geographical races," and "subspecies." In any case the application of these terms in paleontology is bound to be somewhat arbitrary; for, on evolutionary principles, each of the hierarchical systems of categories must grade insensibly into another. In the course of evolution, local groups, isolated by geographical or other barriers, will tend to undergo a gradual genetic diversification until they are sufficiently distinct to justify

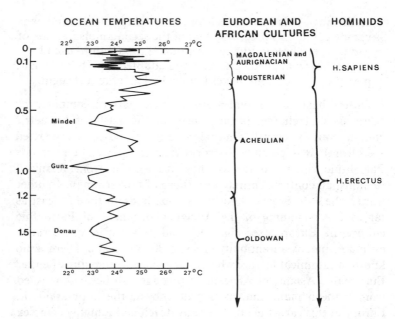

Fig. 4. Probable relationships of Paleolithic cultures and species of *Homo* to warm and cool phases of the Pleistocene. The temperature chart shows the kind of data now being obtained from the study of Foraminifera in deep-sea cores (Hays et al. 1977). For further details, see Oakley (1969).

recognition as different demes or races. With continued isolation the latter acquire a more contrasted genetic differentiation that may be expressed by placing them in separate subspecific categories. A continuation of the same process leads ultimately to the establishment of separate species that are genetically incompatible to the extent that they are no longer freely interbreeding. Thus, in the evolutionary sense, a geographical race is a potential subspecies, and a subspecies is a potential species. Since all the available evidence shows that in the course of progressive differentiation the transition from lower to higher categories in the taxonomic scale is a gradual one, it is not to be expected that these categories can be defined with any precision. If the definition of a subspecies offered by Mayr (1942) is followed, that is, "a geographically localized subdivision of the species which differs genetically and taxonomically from other subdivisions of the species," then the major races of mankind today are properly to be regarded as subspecies (equivalent, that is to say, with subspecies of other vertebrate species). Of course, in recent centuries the factor of geographical isolation of human populations has become grossly disturbed, with the result that interbreeding (particularly along regions of territorial overlap) has produced intergradations between one major race and another and thus has tended to obscure genetic and phenotypic contrasts that must at one time have been more abrupt. A similar process of hybridization between subspecies among animals in the wild state is well recognized by zoologists. For example, instances have been recorded among birds where, after a breakdown of ecological barriers or other isolating mechanisms, two closely related subspecies freely interbreed, with the resulting development of new hybrids. In the course of time such a process may obliterate former subspecific differences; but this, of course, does not invalidate the application of the term "subspecies" to the interbreeding communities while the latter still exist in the main as recognizably distinct groups (Garn 1971).

The determination of the evolutionary differentiation of the major races of *H. sapiens* presents exceptionally difficult problems for the paleontologist. These races may be distinctive enough in the flesh, but the anatomical distinction between them is not reflected to anything like the same degree in the bones. Though they are recognizable by a number of external and rather superficial characters such as skin color, hair texture, and nose form, they are by no means so easily distinguishable by reference to skeletal characters alone—at least not in the case of individual and isolated specimens (which are usually all that paleoanthropology has to offer). It may indeed be possible to identify a skull of a modern Negro, an Australian aboriginal, or a European, in individual cases where the racial characters are exceptionally well marked; but the variation within each group is so great that skulls of each type may be found that are impossible of racial diagnosis. This difficulty may be illustrated by reference to the well-known skulls from the late Paleolithic sites of Grimaldi near Mentone and Chancelade in the Dordogne Valley, on the significance of which there has in the past been considerable controversy.

The Grimaldi skulls had been held by many anthropologists to be definitely Negroid in type, and they were commonly accepted as good evidence of the existence of a Negroid race in Europe during Aurignacian times. Yet others, also from a study of the actual remains, have pronounced categorically that they show no real evidence of Negroid affinities but are simply variants of the "Mediterranean race," somewhat distorted perhaps, that now inhabits southern Europe. The Chancelade skull, of Magdalenian age, had been compared by the famous anatomist Testut with that of an Eskimo, and this diagnosis was accepted by many reputable anthropologists. This conclusion, again, has been vigorously denied by equally distinguished anthropologists, who maintain that it is only a variant of the Cro-Magnon type of *H. sapiens.* Now it is probable that there are no racial types in which the skull characters are more distinctive than Negroes and Eskimos; yet experts fail to agree when faced with single skulls whose claims to these types are in question. If a decision proves so difficult in such cases, it will be realized how much more difficult, or even impossible, it will be to identify, by reference to limited skeletal remains, minor racial groups with less distinctive characters.

If there are difficulties in identifying the skeletal remains of the fully differentiated racial types of today, these are very much enhanced when attempts are made to detect corresponding racial differences in prehistoric remains in which the characteristic racial traits were presumably much less developed. Yet is has been argued by some anthropologists, on the basis of such ancient fossils, that the primary races of mankind had already begun their divergent differentiation as far back as the beginning of Pleistocene times (see p. 88). Though there is no inherent impossibility in such a thesis, the anatomical evidence on which it is based remains entirely inadequate. The shape of the incisor teeth and the presence of exostoses along the alveolar margin of the mandible, for example, are not sufficient to justify the conclusion that Pekin man was the forerunner of the Mongolian races to the exclusion of other racial types. There is likewise no distinctive character in the Rhodesian skull that permits the assumption that Rhodesian man was a proto-Negroid (though in some museums restorations of this fossil man are provided with typical Negroid features and hair of the usual Negroid type).

It must unfortunately be recognized, then, that there is as yet no sound paleontological evidence by which the antiquity of the infraspecific groupings of modern man can be satisfactorily determined. There is certainly no evidence that modern racial differentiation had become fully established before the terminal phase of the Pleistocene period. The local geographical areas where, as the result of temporary isolation, there first occurred the genetic diversification that gave rise to the major races are either unknown or can only be indirectly inferred by a considera-tion of their present-day distribution. No fossil skeleton that is indisputably older than the end of the Pleistocene has yet been discovered that can be certainly identified as of Negroid or Mongolian stock, though the skulls of Australoid type found at Talgai, Keilor, and Mungo have now been assigned a late Pleistocene date. The skeletal remains of Aurignacian and Magdalenian date that have so far been discovered in Europe not only are indubitably those of *H. sapiens* but are actually not distinguishable, on present evidence, from those of modern Europeans. Mention should be made of a number of fossil skulls of modern human type found in Africa, China, and America and dating from the Late Pleistocene. Most of those from Africa have been stated to be of a "Bushmanoid" type; those from America have been compared with the modern American Indians; and a presumed family of three individuals from the Upper Pleistocene of Choukoutien (near Pekin) has been compared by Weidenreich (1939) individually with Mongoloid, Melanesoid, and Eskimoid types. In fact, however, it seems that the racial affinities of all these specimens are really indeterminate on the basis of such scanty material; and it may be questioned whether the opinions of those anthropologists who have described them were not in-fluenced by the geographical locality where they were found.

The question now arises whether *H. sapiens* as a species has a still greater antiquity than the Upper Pleistocene. From time to time in the past, claims for a remote antiquity have been made for skulls and skeletons of modern human type. The famous Galley Hill remains—found as long ago as 1888 in Middle Plestocene gravels of the 100-foot terrace of the Thames—were accepted for many years by some authorities as contemporaneous with the deposits from which they were disinterred. The Ipswich skeleton, found beneath 4 feet of boulder clay, was assumed to antedate the last glaciation of the Ice Age. The Calaveras skull, found 130 feet deep in a gold mine in Table Mountain, Calaveras County, California, was found in gold-bearing strata believed to be some ten million years old. In all these cases (as in a number of other reported discoveries of a similar character) the geological evidence of antiquity was actually quite inadequate; but, even so, it was avidly seized by those who were particularly anxious to bolster up arguments for the remote origin of *H. sapiens*. Such arguments were even used to refute the general conception of human evolution by antievolutionists, who argued that, since human remains of modern type evidently antedated more primitive types that were regarded as ancestral to *H. sapiens*, they upset the morphological sequence that was supposed to provide the essential evidence for human evolution. In the case of the Ipswich skeleton, the initial claim for its antiquity was subsequently shown to have been based on a misinterpretation of the geological evidence, and the skeleton is now regarded as a secondary interment (probably of relatively recent date). The real nature of the Galley Hill skeleton was later determined by the analysis of its flourine content (Oakley and Montagu 1949), from which it is now quite evident that the skeleton is also of no great antiquity—perhaps the remains of a burial in Neolithic times. It is extremely interesting that such an analysis was also made of the Calaveras skull by Thomas Wilson of Harvard University in 1879, but the results, although they showed the skull to be recent and intrusive, were overlooked (Oakley 1964*b*).

Since the method of estimating the relative age of fossil bones by fluorine analysis has proved in recent years to be so valuable, it is appropriate to make further reference to it here. Many years ago it was demonstrated that the fluorine content of fossil bone increases with geologic age. This is because, by a process of ionic interchange, fluorine is slowly taken up from the soil in which the bone is imbedded and becomes fixed in the form of a very stable compound, fluorapatite. The amount of fluorine in a fossil bone thus increases with time and gives an indication of the period over which it has lain in position in a particular geologic deposit. But the amount of fluorine taken up also depends, of course, on the amount of fluorine in the soil. If the soil is rich in fluorine, any bones imbedded in it may become so rapidly saturated that a fluorine analysis is of little use in demonstrating their age. It must be emphasized, therefore, that the analysis permits an estimation only of the *relative* antiquity of fossil material from the same deposit. Thus it does not permit a comparison of the relative (or the absolute) antiquity of fossilized bones derived from different deposits, in which the fluorine content of the soil may vary widely. But in a case where, in the *same* geological deposit, a human skull is found in association with the skeletal remains of extinct mammals of known antiquity, the fluorine test may provide evidence of the utmost importance for determining whether they are contemporaneous (that is to say, whether they were initially placed in the deposit at approximately the same time). If, on the other hand, the human skull represents part of an artificial interment at a much later time, this would at once be demonstrated by its low fluorine content as compared with indigenous fossil bones. The method of fluorine analysis has been developed and applied (particularly by Oakley 1953) in a number of doubtful and disputed discoveries of human remains to which some authorities had attributed great antiquity.

In the case of the Galley Hill skeleton, the fluorine content was found to average 0.34 percent. This compares with a range of 1.7-2.8 percent in Middle Pleistocene bones from the same region, 0.9-1.4 percent for Upper Pleistocene material, and 0.05-0.3 percent for Holocene bones. The differences are sufficiently marked to justify the firm conclusion that the Galley Hill skeleton "was not indigenous to the Middle Pleistocene gravels in which it lay, but a burial of later date—prehistoric, but probably post-Pleistocene" (Oakley and Montagu 1949). Thus, after more than fifty years of argument this way and that on the basis of inadequate geological evidence, the question has now been finally settled by fluorine analysis, and no better example could be adduced to illustrate the value of this crucial test.[2]

Some methods that are now available for estimating with considerable accuracy the *absolute* antiquity of fossils are based on the fact that certain radioactive substances often to be found in fossil-bearing strata undergo a gradual disintegration at a known rate unaffected by normal climatic changes. One such method makes use of the disintegration of the element uranium to form helium and lead. The rate of decay in this case is very slow, one million grams of uranium producing 1/7600 gram of lead every year, but by chemical estimation of helium and lead in appropriate minerals the lapse of time since the first formation of the latter can be calculated. This uranium-lead method is applicable only to very ancient deposits, but it has determined the age of certain deposits of the Miocene period (estimated at approximately twenty million years) that has a special interest for hominid evolution, for it was during this period that large apes of a very generalized type abounded in many parts of the Old World, so generalized that some of them may well have provided the ancestral basis for the subsequent emergence of the earliest hominids.

Another method more recently developed is the potassium-argon method, depending on the fact that naturally occurring potassium contains 0.01 percent of a radioactive isotope that on decay forms calcium and argon and has a half-life of thirteen hundred million years. This has given with more precision the overall age of Miocene deposits extending back from five million years ago to twenty-five million years, that of the subsequent Pliocene period extending from about one and a half to five million years, and the beginning of the following Pleistocene period an antiquity of about one and a half million years. And it was during the Pleistocene period that hominid evolution gradually proceeded toward the final appearance of the genus *Homo* and the species *Homo sapiens*. This method, however, cannot be used with much accuracy to determine ages of less than about half a million years.

Last, we may refer to the important radioactive carbon method. This depends on the fact that, during life, organisms contain a constant amount of radioactive carbon (carbon 14) irrespective of geographical distribution or other environmental factors, and that on death the carbon 14 disintegrates at a rate of one-half its amount in 5,730 years; thus an estimation of its quantity in organic material that is suspected to be ancient will provide good evidence of its absolute antiquity. But, because of its short half-life and the technical difficulties of estimating minute quantities, the limit of dating by this method is not more than 50,000 years. The accuracy of dating by estimating the products of decay of any radioactive element of course demands very careful attention to a number of technical details, but on the whole the results so far obtained have been remarkably consistent.

Apart altogether from those fossilized remains of *H. sapiens* whose antiquity must remain entirely dubious because they are so inadequately documented,[3] there are now available a small number of specimens apparently of this species that can be assigned with reasonable assurance to the Middle Pleistocene or the early part of the Upper Pleistocene. But, before considering these, we should give attention to what at one time was regarded by some authorities as a distinct species of *Homo* (*H. neanderthalensis*), which occupied many parts of the Old World during the first half of the Upper Pleistocene.

Neandertal Man and His Contemporaries

Since the discovery of a skull cap and portions of the limb skeleton in the Neandertal cave (near Düsseldorf) in 1856, "Neandertal man" has now become recognized by most anthropologists as representing a distinct group of the species *Homo sapiens* that probably became differentiated in the latter part of the Middle Pleistocene period but did not survive the end of the

Pleistocene. Following on the earlier discovery, other specimens of this race have been described from a number of important sites in Europe—for example, Gibraltar, La Chapelle-aux-Saints, La Quina, Spy, La Ferrassie, Monte Circeo, Le Moustier, La Naulette, and Jersey. Fragmentary Neandertal remains are known from many other sites, but it will be possible here and throughout this book to describe only the most complete and significant specimens.[4] All these fossil specimens were almost certainly contemporaneous with the later part of the Mousterian period of Paleolithic culture and were representatives of a population that lived during the first phase of the last glaciation (Würm I) of the Ice Age. This population is sometimes referred to as West European or "classic Neandertal man" or "Mousterian man" (though actually it is characteristic only of the later Mousterian period). The cranial characters and such elements of the postcranial skeleton as are available have been studied in considerable detail by a number of competent anatomists, and there is a general consensus that the differences they show from *H. sapiens* are consistent enough to justify at least some degree of taxonomic distinction (see fig. 5). Today they are usually considered to represent a distinct subspecies of *Homo sapiens*; *Homo sapiens neanderthalensis*. The diagnostic characters of the group may be defined as follows.

> *Homo sapiens neanderthalensis*—the skull is distinguished by an exaggerated development of a massive supraorbital torus, forming an uninterrupted shelf of bone overhanging the orbits (with complete fusion of the ciliary and orbital elements); absence of a vertical forehead; marked flattening of the cranial vault (platycephaly); relatively high position of the external occipital protuberance and the development (usually) of a strong occipital torus; a massive development of the naso-maxillary region of the facial skeleton, with an inflated appearance of the maxillary wall; a heavy mandible, lacking a chin eminence; a pronounced tendency of the molar teeth to taurodontism (enlargement of the pulp cavity, with fusion of the roots);[5] a relatively wide sphenoidal angle of the cranial base (about 130°; see fig. 10, p. 85); angular contour of the occiput; certain morphological details of the ear region of the

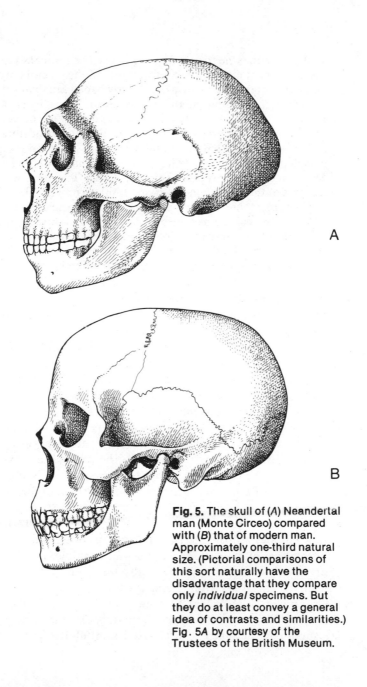

Fig. 5. The skull of (*A*) Neandertal man (Monte Circeo) compared with (*B*) that of modern man. Approximately one-third natural size. (Pictorial comparisons of this sort naturally have the disadvantage that they compare only *individual* specimens. But they do at least convey a general idea of contrasts and similarities.) Fig. 5*A* by courtesy of the Trustees of the British Museum.

skull (including the rounded or transversely elliptical shape of the auditory aperture, the conformation of the mastoid process and of the mandibular fossa); a slightly backward disposition of the foramen magnum; and a large cranial capacity (1,300–1,600 cc).[6] The limb skeleton is characterized by its considerable robusticity but is nevertheless not morphologically or functionally different from that of modern man. The total morphological pattern of the lower limbs suggests that squatting was a common position of repose (Trinkaus 1975). The morphological features of the pubic bone (Stewart 1960, Trinkaus 1976) are distinct, together with certain of the morphological details of the talus and calcaneus bones of the ankle. In addition, the vertebrae of the cervical region of the spine in some cases show a striking development of the spinous processes, but it does not exceed the extreme limits of variation in modern man.

All these morphological factors are not known for certain to have been present in all the Mousterian remains listed earlier, for some of the latter are represented only by fragmentary skulls. Here we have exemplified one of the inherent difficulties of much paleontological work—the attempt to define a taxonomic group on the basis of skeletal remains that are rarely complete (or anything like complete) in *individual* specimens. It is a difficulty, however, that constantly must be met as best may be (even if only on a provisional basis) by reference to such material as is available to date. Fortunately, in the present instance the total morphological pattern presented by the cranial and skeletal characters just enumerated is rather distinctive; it is not to be found in any of the races of modern man. Also, very fortunately, the cranial characters have been analyzed and assessed on a statistical basis in a systematic study by Morant (1927). This author emphasized the metrical features in which the skull differs from that of modern *H. sapiens*—for example, the skulls are "particularly characterized by the absolutely and relatively large size of the facial skeleton"; and a comparison of the relevant measurements is sufficient to show that the form of the facial skeleton "dissociates the type" from modern man; nearly all measurements designed to assess the sagittal flattening of the cranial vault "relegate the Mousterian skulls to positions which are entirely outside the inter-racial distributions for modern man"; the axis of the foramen magnum is more deflected from the vertical (owing to a rotation of the occipital bone) than in modern races; the skulls are "distinguished from all modern types by having a greater transverse flattening of the vault, more vertical walls and a height that is peculiarly small in proportion to the breadth"; and, as regards the breadth-length indices of the separate frontal, parietal, and occipital bones, "some fall entirely outside the interracial range for modern skulls."

Possibly the most important functional component of the morphological pattern of the Neandertal skull is masticatory. The relatively large jaws and teeth require a large and well-built mandible, facial skeleton, frontal bone, and associated musculature. With this heavy masticatory apparatus, the skull, balancing on the occipital condyles, requires increased nuchal musculature to support it and better leverage for these muscles. Thus the development of the masticatory apparatus not only may account for the heavy face and brow ridges that transmit the stress in chewing, but may also explain the long, low skull, well-developed occipital torus, and angular contour of the occiput—the Neandertal "bun."

A biometrical study has been made of the Neandertal femur by Twiesselman (1961). He emphasizes the remarkable homogeneity of the thigh bone (in the Western Neandertal population of the last glacial period) and the differences in certain indices and dimensions from those of modern *Homo sapiens*. This study, together with more recent multivariate analyses (Stringer 1974*b*), thus extends and confirms the conclusions drawn by Morant from his own study of cranial characters. The Mousterian fossil remains that are assignable to Neandertal man on the basis of these metrical and morphological criteria seem to have represented a comparatively isolated group occupying for the most part a restricted geographical area for a limited period of time during the first (and most severe) phase of the last glaciation. Indeed, it may well have been the isolation caused by the rigorous climatic conditions that actually led to the differentiation of the group;

for, as Howell (1952) has noted, "under this environment selection would be severe, chance for genetic drift at an optimum, and opportunities for migration reduced to a minimum." It is to the representatives of this particular group that the term "Neandertal man" was originally applied. But the term has also been somewhat loosely applied to human remains from other parts of the Old World of the same age and of still greater antiquity (contemporaneous, that is to say, with the interglacial phase preceding the Würm glaciation or even earlier), because in some of them the supraorbital ridges and other features of the robust Neandertal cranium are rather strongly developed and thus tend in appearance to approximate those that form one of the characteristics of later Mousterian man. However, the supraorbital ridges in the pre-Mousterian or early Mousterian representatives of *Homo* usually correspond more closely with those of other races of *H. sapiens* in the tendency they show to divide into medial (ciliary) and lateral (orbital) elements separated by a shallow groove (sulcus supraorbitalis). This twofold composition of the supraorbital ridges in *H. sapiens* is well recognized in those modern races in which the ridges are strongly developed; and Howell rightly points out that in this feature some of the earlier skulls are very similar to Australian and Tasmanian skulls. In West European Neandertal man, on the other hand, the ridges are specialized to form a continuous, uninterrupted torus.

A number of human fossils have been found in other parts of the world, beyond central Europe and the Pyrenees, that are roughly contemporary with West European Neandertal man and show a reasonably close morphological relationship with the classic population. The most important of these will now be briefly described.

The Mount Carmel and Galilee Skeletons

The important skeletal material found in Israel since 1925 comes mainly from the areas of Mount Carmel and Galilee and was excavated from a series of caves. The material falls into two groups, the older, which bears considerable similarity to the Neandertal people of Europe, and the younger group, which is very much more modern in appearance. Of the older group, dated from the early part of the Würm glaciation, the most famous discovery comes from the cave of Tabun, from which we have an almost complete skeleton of a woman and a male mandible. From the caves of Amud and Zuttiyeh in Galilee we have a skeleton (with other fragments) and a skull respectively. This older group, probably dating from over 50,000 years BP, all show characters in common with the West European Neandertal group, though there are some minor differences: the Israeli people tend to be less heavily built and to carry more rounded skulls. The mandible found at Tabun has an incipient mental eminence, but this feature is also occasionally seen among Neandertal man. On the whole these people have Neandertal features, but they are less strongly expressed than in their West European contemporaries.

The second group consists of some twenty skeletons from the caves of Skhūl and Djebel Kafzeh. They show a much more modern morphological pattern in the skull, and the skeleton is quite indistinguishable from that of modern man. The skulls still show a considerable development of the supraorbital ridges, but in the rounded forehead and the height of the cranial vault, some closely approximate modern man; and this resemblance is enhanced by the rounded contour of the occiput, the strongly developed mastoid process, the smaller sphenoidal angle of the cranial base, the development of a distinct mental prominence on most of the mandibles, the moderate size of the facial skeleton (with a hollowing of the maxillary wall), and the slender, straight-shafted, limb bones. To summarize, the Tabun remains approximate more closely the Neandertal type, while those from Skhūl show a general similarity to *Homo sapiens* of the Upper Paleolithic. It may be suggested, therefore, that the former were replaced by the latter about 40,000 BP during the shifting movements of populations in the Near East at this time. Taken in conjunction with the evidence presented by other examples of pre-Mousterian man, some of these populations represent a

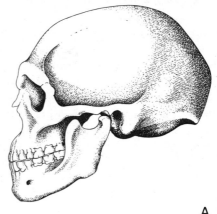

A

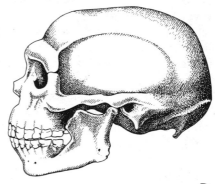

D

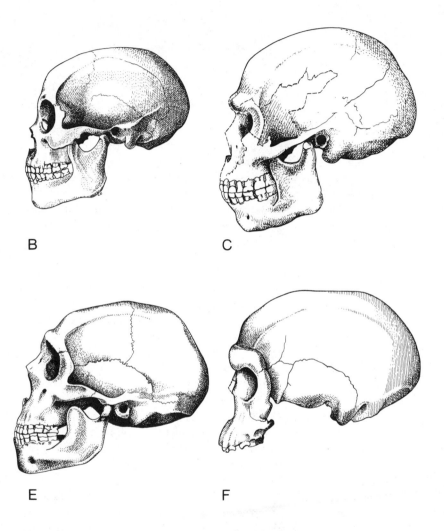

B

C

E

F

Fig. 6. Skulls of Australian aboriginal (*A*) and African pygmy (*B*); Skhūl 5 (*C*) and Tabun 1 (*D*) from Mount Carmel, Israel; Shanidar 1 (*E*) and Djebel Irhoud 1 (*F*). Approximately one-quarter natural size.

transitional stage leading from pre-Mousterian *H. sapiens* to the later *H. sapiens* of modern type.

Solo Man

The "Neandertaloid" remains from Java were found at Ngandong in terrace deposits related to the River Solo (Oppenoorth 1932). The geological and paleontological evidence suggests that they are of Upper Pleistocene date, the latter probably corresponding to the last glaciation (von Koenigswald 1949). The remains, discovered between 1931 and 1933, consist of eleven skulls (all lacking the facial skeleton) and two incomplete tibiae. A detailed and well-illustrated description of this material by Weidenreich was published in 1951, but, owing to his sudden illness and death, this is not complete. The skulls show very interesting resemblances to the Rhodesian skull, with marked platycephaly, a powerful development of the supraorbital tori, and unusually thick cranial walls (fig. 7). Except for the cranial capacity (which ranges from 1,150 to 1,300 cc), and certain unusual features of the foramen magnum (e.g., its length and the upward deflection of its posterior part), the skulls also approximate West European Neandertal man. On the other hand, the two tibiae are slender and straight-shafted and appear to show no difference from the tibia of modern man. Here the lack of fossil data makes it impossible to arrive at any firm conclusion; the fact is that much more of the limb skeleton of the Ngandong population of Java and of later Mousterian man elsewhere is needed for a satisfactory comparative study. Oppenoorth (1932), who first described the Ngandong skulls, allocated them to a new genus, *Javanthropus*, but he later retracted this opinion and gave them the name *H. soloensis*. In this he was followed by Weidenreich, though neither of them attempted to formulate a diagnosis of this new species. Probably it is wiser to consider them geographical races or subspecies of *Homo sapiens* of equivalent standing to the West European Neandertal race. It is convenient to refer to them colloquially as "Solo man."

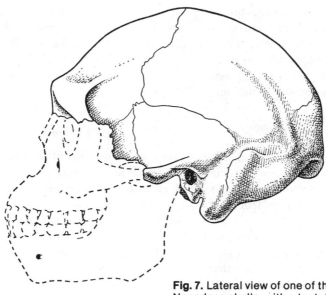

Fig. 7. Lateral view of one of the
Ngandong skulls, with a tentative
reconstruction of the facial
skeleton (after Weidenreich).
Approximately one-third natural
size.

The Shanidar Skeletons

Between 1953 and 1960, Ralph Solecki excavated seven skeletons
from the Shanidar cave in Northern Iraq. The fossil fauna was
indistinguishable from the local extant fauna, but the industry
was Mousterian (Solecki 1960). The fossil hominids have been
dated radiometrically on the basis of charcoal in associated
hearths to lie between 47,000 and 60,000 BP. The seven skeletons,
one of which was a flexed burial and three of which were
accidental deaths from rock-falls, are clearly very similar to the
classic West European Neandertal people (fig. 6). They do show

some differences, however; they are taller and less robust, with more rounded skulls, especially in the occipital region, and less prognathous jaws. These people seem close to the Tabun skeleton in their overall morphology (p. 67). However, of all the Asian finds, this group probably lies closest to the West European Neandertal population.

The Skull from Mapa, China

One important skull of about this period was found in a cave at Mapa, Kwantung Province, in 1958. Its age is rather uncertain: it could be contemporaneous with either Solo or Neandertal man. It consists of the frontal, parietal, and nasal bones only, but the heavy brow ridges and retreating forehead are well preserved. Chinese scholars emphasize its close morphological relationship with Solo man, though it also appears to have much in common with Shanidar and other "Neandertaloid" specimens.

"Neandertaloid" Remains from Africa

The Rhodesian fossils were found in 1921 in the course of opencast mining of lead and zinc ores at Broken Hill in Zambia. They consist of a skull (almost complete except for the mandible), a maxillary fragment, a sacrum, and portions of the pelvis, humerus, femur, and tibia (Pycraft et al. 1928). The skull is of unusually massive appearance and markedly platycephalic, with a very strongly developed supraorbital torus, retreating forehead, a strong occipital torus, and a large inflated maxilla lacking a canine fossa. The palate is unusually large, having an area of 41 cm^2 (as compared with an average of about 25 cm^2 in modern English skulls). In these characters the Rhodesian skull undoubtedly shows a close resemblance to Neandertal man and is therefore sometimes described as "Neandertaloid." Yet there are certain differences, whose significance is not easy to assess. Morant (1928) has shown that the skull can be clearly distin-

guished in a number of independent metrical characters, such as
the relative position of the maximal parietal breadth, the basal
and facial lengths, and the absence of a backward rotation of the
occipital bone; and in these characters it actually approaches
somewhat more nearly to modern man. In yet other features it
shows a greater divergence from modern man. In general,
however, Morant concludes that Rhodesian man and Neandertal
man—in the limited number of purely metrical characters of the
skull he used for comparison—appear to be more closely related
to each other than either is to modern man.

The evidence of the skull of Rhodesian man must now be
considered in relation to the other remains found at Broken Hill.
In the first place, the separate maxillary fragment (described by
Wells 1950) is considerably smaller than the corresponding region
of the type skull, the dimensions of the palate coming very close to
the maximum dimensions recorded for that of the Australian
aboriginal and African Negro. It also has a definite canine fossa,
the modeling of this part of the maxilla "being essentially as in
modern human skulls." Second, the pelvis,[7] sacrum, and limb
bones are entirely similar to those of modern man and show none
of the features commonly regarded as characteristic of Neandertal man. The question naturally arises whether the maxillary
fragment and the portions of the postcranial skeleton were
actually contemporaneous with the skull. However, recent studies
by Oakley (1950), based on a spectrographic analysis of the lead
and zinc content of all the remains, have led him to conclude that
"there is little reason for doubting that at least all the human
bones recovered from the cave in 1921–22 form a contemporary
group"; and their size and robustness are also consistent with the
inference that they are all referable to the same type as the skull.
The artifacts, which belong to the Proto-Stillbay phase of the
South African Paleolithic period, have been studied in detail by
Clark (1950), and this archeological evidence indicates an antiquity of Upper Pleistocene (ca. 40,000 BP).

A second skull, almost identical in contour and morphological
details with the Rhodesian specimen, was found in 1953 at Hope-

field near Saldanha Bay, ninety miles north of Cape Town (Singer 1954). The site of the discovery contained stone instruments of the Chelles-Acheul Paleolithic culture. The associated remains of fossil mammals again indicate an Upper Pleistocene period. Thus it appears that this type of early man was widely distributed in Africa at this time. A small fragment of the lower jaw found in association with the calvaria of "Saldanha man" resembles in shape and dimensions the corresponding part of the Heidelberg mandible (see below, p. 115).

The Omo Skulls

In 1967 Richard Leakey collected fragments of three skulls near the Omo River in southeast Ethiopia. They have been dated only approximately to 100,000 BP and are associated with no clearly defined industry or fauna. The two calvaria are strikingly different in morphology. Omo 1 has a more rounded cranium and has been compared with Skhūl V. It is associated with a mandible with chin and a number of skeletal remains. Omo 2 and the Omo 3 fragment are much more robust and comparable to the Solo skulls in some features, to Broken Hill in others (Day 1969). These finds, however, should not be considered exact contemporaries: their age is uncertain, and they were found 3 km apart. Their association nevertheless reminds us of the kind of variation we may expect to find at one period and in one region.

The Jebel Irhoud People

Two well-preserved adult skulls—a cranium and a calvaria— together with a juvenile mandible were excavated from a mine near Safi, Morocco, between 1961 and 1963. They are of approximately Early Würm age (though there is no direct evidence of the Ice Age in North Africa) and so are roughly contemporary with Neandertal man. They were accompanied by a Levallois-Mousterian industry. Again, these skulls bear Neandertal features but have slightly rounder craniums than the classic population. However, Irhoud 1 is strongly prognathous and lacks a canine fossa, which implies a heavily built dentition (fig. 6) (Ennouchi 1962). Of skulls found in Africa, these are probably most closely allied to the European Neandertal people.

Early *Homo sapiens*: Riss-Würm

Beside the group of distinctive Neandertal specimens from West Europe, a number of other human fossils are now known from the last interglacial period (Riss-Würm) and the preceding interglacial phase (Mindel-Riss) in West Europe (see fig. 4). At present the fossil evidence suggests that man did not survive the Riss and earlier glaciations in central and northern Europe but moved north at the beginning of each interglacial from the warmer, more southern parts of the Continent. As far as we know the first humans to survive the arctic conditions of a glacial epoch were the Neandertals of West Europe.

The people of these earlier two interglacial periods are not strikingly dissimilar to modern man, and they are therefore usually considered the earliest members of the species *Homo sapiens*. We shall now review these fossils—first the more recent ones from the Riss-Würm, then the much more ancient group from the Mindel-Riss.

The Ehringsdorf Skull

The Ehringsdorf skull was found in 1925 at a depth of 54 feet in travertine deposits near Weimar, in association with fossil remains of *Elephas antiquus, Dicerorhinus merckii, Bos,* and *Equus* and with fossil evidence of a temperate forest flora. The consensus assigns the skull to the last interglacial period. The skull consists of the braincase only, and this has had to be reconstructed from a number of isolated fragments. However, there is little doubt that Weidenreich's reconstruction (1928) gives a reasonably accurate representation of the calvaria (though it does not permit complete accuracy of measurement of overall dimensions). The supraorbital ridges are heavily built and in this respect (like those of the Steinheim skull) show some degree of approximation to the massive supraorbital torus of Neandertal man. On the other hand, the frontal region is filled out by a pronounced convexity to form a vertical forehead, the cranial vault is relatively high, and there is a well-developed pyramidal mastoid process. The cranial capacity has been estimated at about 1,450 cc, but, in view of the damaged condition of the skull, this can be regarded as only a very rough approximation. If the Ehringsdorf skull is considered as a morphological problem in itself (with the necessary qualification that the facial skeleton and the postcranial skeleton, if known, might lead to other conclusions), it can certainly be stated that the calvaria approximates modern man more closely than does Neandertal man.

The Krapina Skeletal Remains

The Krapina skeletal remains were found in the floor of a rock shelter in northern Croatia, and the first results of their study were reported in 1901 by Gorjanovič-Kramberger (1906). The basal strata (in which the human remains were found) are fluviatile, but the floor of the cave today is 25 meters above the level of the present river. In the same strata were found the remains of a warm interglacial fauna, including *Dicerorhinus merckii*. It is now generally agreed that the deposit belongs to the last interglacial period and is probably contemporary with the Ehringsdorf site. The human remains are numerous but fragmentary; they include portions of several skulls, many teeth, and parts of the postcranial skeleton. All the adult skulls show a strong development of the supraorbital ridges, and in a number of other features, such as the sloping forehead, powerful jaws, and small mastoid process, some of them approximate Neandertal man of later Mousterian date. On the other hand, the skulls show a considerable degree of variation, for in some the frontal region is closely similar to that of modern man, and the skull vault is relatively high. The limb bones, also, are not distinguishable from modern man. The Krapina population, like the people from Israel, shows characters intermediate between the extreme

West European Neandertals and modern man. We shall discuss the status of these intermediate populations at the end of this chapter.

The Saccopastore Skulls

The Saccopastore skulls were found (in 1929 and 1935) in a river deposit on the bank of a small tributary of the Tiber, at depths of 6 and 3 meters respectively. In the same deposit were found remains of *Elephas antiquus* and *Hippopotamus*, and there can be little doubt that the human remains date back to the last interglacial period. The skulls show a strong development of the supraorbital ridges, some degree of platycephaly, and a rather massive maxilla with reduced canine fossae. In these features they show some of the characteristics of Neandertal man. By contrast, however, the cranial capacities are relatively low (ca. 1,200 and 1,300 cc), the foramen magnum is advanced in position as in modern man, the sphenoidal angle of the cranial base is small (101°–105°), the occiput is well rounded, and the dental arcade is very similar to that of modern man (with a marked reduction of the last molar tooth). Sergi (who has published detailed descriptions of the Saccopastore material) expresses the opinion that the remains represent an initial phase in a progressive development that later led to the specialized West European Neandertal race (Sergi 1943, 1944).

The Fontéchevade Skulls

In 1947, portions of two skulls were found in situ by Mlle G. Henri-Martin at a depth of 2.6 meters in a cave at Fontéchevade in southern France. The stratum containing these remains lies below deposits of Mousterian date and is separated from the latter by a layer of stalagmite that had been undisturbed. It contained implements of the Tayacian type (Lower Paleolithic), and a warm-temperate fauna characterized by *Dicerorhinus merckii*, *Dama*, and *Testudo graeca*. The archeological and faunal data are consistent with an antiquity corresponding to the last interglacial (Riss-Würm) phase of the Ice Age. Finally, the fluorine test showed the skull bones to contain 0.4–0.5 percent fluorine, as compared with a range of 0.5–0.9 percent for mammalian bones of Tayacian date and 0.1 percent or less for human and mammalian bones from the superimposed Aurignacian level. As in the case of the Swanscombe skull bones, therefore, the evidence for the antiquity of the Fontéchevade skulls seems to be well assured. However, following a number of studies, it appears that owing to the fragmentary nature of the specimens their morphological status must remain uncertain. The fragmentary frontal bone lacks a supraorbital torus but may belong to a juvenile individual. The skull cap is very badly damaged and distorted but in its thickness can be equated with the much older Swanscombe bones (see below).

The Quinzano Occipital Bone

The Quinzano occipital bone was found in cave deposits near Verona in 1938. The main interest of this fragment lies in its close resemblance to the occipital bone of the Swanscombe skull (see below), as shown in its thickness, great biasterionic width, and the considerable asymmetry of the venous grooves. It was found in a Pleistocene stratum that also yielded *Elephas trogontherii* and *Megaceros* and implements of an archaic Mousterian type, recalling in their technique the Clactonian and Levallosian cultures. This fossil bone has been described by Battaglia (1948), who refers it to the Riss-Würm interglacial phase (at the latest)— that is, to the same antiquity as the remains of the Ehringsdorf and Fontéchevade skulls.

Homo sapiens from the Riss and Mindel-Riss

The Steinheim Skull

In 1933 the Steinheim skull (fig. 8) was found near a tributary of the River Weimar (30 km north of Stuttgart) in interglacial gravels also containing remains of *Elephas antiquus* and *Dicerorhinus merckii*. These gravels are believed to date from the end of

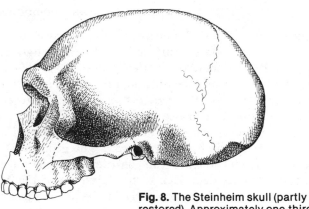

Fig. 8. The Steinheim skull (partly restored). Approximately one-third natural size. By courtesy of the Trustees of the British Museum.

the much more ancient Mindel-Riss interglacial period (ca. 250,000–200,000 BP). The skull is estimated to have a cranial capacity of about 1,100 cc and shows only a moderate degree of platycephaly. It also possesses strongly developed supraorbital ridges, which appear to adumbrate the massive ridges so characteristic of the extreme Neandertal type of skull of later Mousterian times (Weinert 1936). In neither of these characters, however, does it diverge to such a degree from modern man as to warrant placing it with the Neandertal group. The ample development of the forehead region, the rounded form of the occiput (with a low and rather feebly marked external occipital protuberance), the contour in horizontal section of the nasomaxillary region (with a hollowing of the maxillary wall), the relatively moderate dimensions of the facial skeleton, and the total morphological pattern presented by the mastoid process, tympanic plate, and mandibular fossa compose a combination of morphological features that conforms entirely with modern man and contrasts with the skulls of West European Neandertal man. The somewhat low cranial capacity also comes well within the range of modern man, and in this feature again the Steinheim skull does not conform with the characteristically large-brained Neandertal man.

The Swanscombe Skull Bones

The Swanscombe skull bones were first discovered in 1935 and 1936, 24 feet below the surface in well-stratified gravels forming part of the 100-foot terrace of the Thames (Ovey 1964). These are definitely interglacial deposits, and the associated fauna includes *Elephas antiquus, Dicerorhinus megarhinus, Megaceros,* and *Dama clactonia.* Also associated with the skull bones were found flint implements that can with certainty be assigned to the early Middle Acheulian hand-ax industry. The geological, archeological, and faunal evidence is all consistent with the conclusion that the skull bones date back to the late part of the Mindel-Riss interglacial period or very early in the Riss. Finally, this has received some confirmation from a fluorine analysis of the bones (Oakley 1953), for the latter have been found to contain 2 percent fluorine which equates well with the percentage found in the bones of extinct mammals indigenous to the Middle Pleistocene terrace deposits at Swanscombe. Taking all the evidence into account, it may be affirmed that in no other example of Paleolithic man is the relative dating more completely attested than it is in the case of the Swanscombe bones.

After a further discovery in 1955, the Swanscombe "skull" (as it is sometimes called) now consists of the occipital and the two parietal bones (Ovey 1964). But these are extremely well preserved and articulate together perfectly. Since the sutures still remain open, the bones evidently belong to a young individual probably twenty to twenty-five years old. The cranial capacity (as inferred by comparative studies of modern skulls whose occipital and parietal bones show similar dimensions and curvatures) was probably about 1,325 cc. The basibregmatic height and maximum biparietal width are indeed rather greater than the means of the corresponding measurements of female British skulls. The inclination of the plane of the foramen magnum shows nothing exceptional. The occipital bone is rather unusually broad, but even this character falls within the range of variation of Recent European skulls. The endocranial cast provides evidence of a richly convoluted brain, differing in no observable manner from that of modern man. Four features of the skull bones distinguish them from modern man—their thickness (particularly in certain regions, such as the anteroinferior angle of the parietal bone and the cerebellar fossae of the occipital), and the excavation (in so young an individual) of the rostral surface of the basioccipital by a backward extension of the sphenoidal air sinus. In addition, the form of the occipital torus and presence of the masto-occipital ridge show similarities to the equivalent structures in some Neandertal skulls.

A recent multivariate statistical study (Ovey 1964) shows clearly that the Swanscombe "skull" is distinct from both modern and Neandertal crania and lies close to the Steinheim skull and Skhūl V. The Steinheim skull is broadly of similar age, unlike that from Skhūl, and therefore is very likely to represent a second member of the same population. We do not have the face of the Swanscombe "skull," but, were we to find it, it might be expected to differ little from that of Steinheim.

A recent discovery (1976) from Biache Saint-Vaast, near Calais, France, may also be a member of this population, but a full description is awaited.

The Arago Fossils

The relatively recent discoveries from the Caune d'Arago, a cave in the foothills of the Pyrenees (1971), have added greatly to our knowledge of Mindel-Riss hominids (Lumley and Lumley 1973). The cave deposits from which these remains were excavated are dated from the very beginning of the Riss glacial period on both faunal and stratigraphic grounds. The fauna included such cold-loving species as *Ursus spelaeus*, *Rhinoceros mercki*, and *Rangifer tarandus*. (There are at present no radiometric techniques available to establish the absolute age of these Middle Pleistocene sites.) The industry is Tayacian. The occupation levels are succeeded by sterile deposits indicating extreme cold. The human fossils include a face and frontal bone (A21) together with fourteen isolated teeth, phalanges, parts of a parietal bone, and one and one-half mandibles. The frontal and face bear a considerable resemblance to the Steinheim skull, especially in the form of the supraorbital torus, the conformation of the orbits, and the marked postorbital constriction. (There are no bones in common with Swanscombe.) The Arago specimen is, on the other hand, larger and much more robust; the maxilla has no canine fossa and shows considerable prognathism, and this links it with Neandertal man and even *Homo erectus* (Choukoutien, Terni-fine, Mauer—see chap. 3). The large mandibles are also reminiscent of *Homo erectus*. They suggest great sexual dimorphism, being very different in size: one (A13) is closer to Mauer, the second (A2) is closer to Montmaurin (see below). This suggests that much (but clearly not all) of the variation in size and robusticity that we find among the Middle and Upper Pleistocene hominid material is probably due to individual variability and sexual dimorphism, as well as to racial differentiation.

Fragments from Sidi Abderrahman and Rabat (Morocco)

Two fragments of a mandible from the Littorina Cave, Sidi Abderrahman, Casablanca, northwest Africa, are considered to be of early Riss age. They are associated with a Middle Acheulian industry. The mandible lies morphologically between the Neandertal people of West Europe and the much earlier *Homo erectus* fossils from the Algerian site of Ternifine. The Rabat mandible, from the same area but of somewhat later date, falls into the same group; both have many features in common with the slightly earlier Arago mandibles. The importance of these fossils is that they demonstrate that early *sapiens* populations were established in northwest Africa during the Riss glaciation when survival north of the Pyrenees (at least) was not possible for *Homo sapiens*. It seems quite possible that populations such as these gave rise to the later Riss-Würm and Neandertal populations of West Europe.

Résumé of the Relationships
of Early *Homo sapiens*

The problem of "Neandertal man" has been the cause of much controversy for many years. As already noted, the term was in the first instance applied to the remains of the later Mousterian population in Europe that was contemporary with the first part of the last glaciation of Pleistocene times. Among the several distinctive features of the skull in this population is the enormous development of the supraorbital ridges, which are united to form a continuous torus. When skulls of earlier date were later discovered, the term "Neandertal" was also commonly applied to them, simply because they also display a strong development of the supraorbital ridges. But, as we have seen, they do not consistently present the combination of other cranial and skeletal characters that have come to be regarded as forming a total morphological pattern distinctive of West European Neandertal man. A full understanding of the relationships of all these specimens can only be finally made possible by the tabulation of the relative and absolute chronology of the Mousterian and pre-Mousterian remains and by a consideration of their morphological characters. Zeuner's study (1940) proved of considerable importance at the time it was published because he based his relative chronology entirely on the evidence provided by geological stratification, archeological data, and faunal associations, without reference to the morphological characters of the remains themselves. Thus he avoided the bias of some earlier workers, who allowed their assessment of relative antiquity to be influenced by preconceived notions that Neandertal man was essentially a primitive type of Paleolithic man presumed to be directly antecedent to *H. sapiens*. On the other hand, Howell (1951, 1952), making use of Zeuner's chronological tabulation, attempted to elucidate the relationships of the various fossil remains by making it clear that the later, or "classic Neandertal," fossils almost certainly represent a peripheral development during isolation in the initial phases of the last glaciation. The earlier types that inhabited Europe during the last interglacial

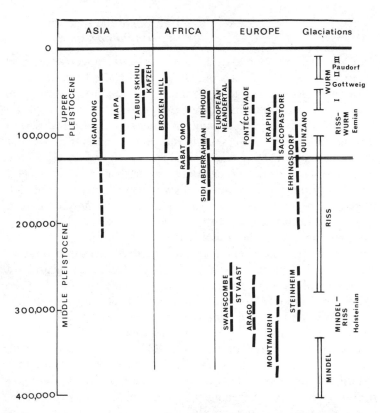

Fig. 9. Ranges of uncertainty in the ages of important *H. sapiens* fossils.

period may be conventionally termed "generalized Neandertals." The major characteristics of these populations are the strong development of the supraorbital ridges and the great thickness of the cranial walls. However, in other features (so far as these can be determined in rather fragmentary specimens) they show a fairly close resemblance to modern man (fig. 10).

It has to be admitted, of course, that the paleontological data regarding the origin and evolution of *Homo sapiens* are still scanty. But, on the basis of the morphological and chronological

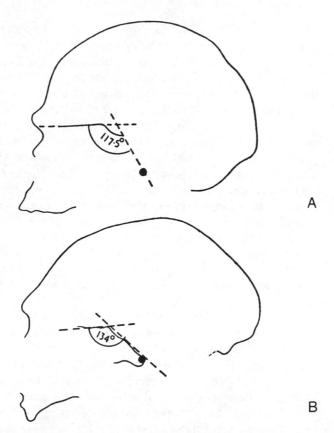

Fig. 10. Midsaggital craniograms of (*A*) an early or generalized "Neandertaloid" skull (Skhūl) and (*B*) a "classic" Neandertal (Monte Circeo), to illustrate the differences in the contours and the flexion of the basis cranii as shown by the sphenoidal angle. In modern man the angle has a mean value of 110°. After Howell.

evidence available at present, the most reasonable interpretation is that a primitive type of *H. sapiens* retaining strongly developed brow ridges came into existence during the Middle Pleistocene (ca. 300,000 BP), presumably from an earlier small-brained type represented by the *Homo erectus* stage of hominid evolution; that these early *H. sapiens* spread widely throughout the Old World, adapting to the differing climates and ecosystems and producing a variety of different races. In other words, it may be surmized that the progressive development of the brain in the *Homo erectus* group of hominids led eventually to the appearance of a type of *Homo* that, on morphological evidence, and in spite of the retention of some primitive characters of the skull, was not specifically distinct from modern man. In some races, the continued expansion of the brain was associated with a more exaggerated development of a supraorbital torus and of the jaws and palate, the appearance of certain specializations of the skull and teeth, and a more robust limb skeleton. In other races a progressive enlargement of the brain was associated with a recession of the jaws and the supraorbital ridges, a diminution in the size of the teeth, the construction of a more evenly rounded cranium with a vertical forehead, and the refinement of the limb characters already developed much earlier in *H. erectus*. These more gracile races evidently led to the modern races of *H. sapiens*.

As we have said, Howell pointed out that the extreme Neandertals of western Europe were probably cut off as geographical isolates by glacial and periglacial regions for several thousands of years, so that they could have had little exchange of genes with hominid populations elsewhere. There would be nothing surprising in this, for it is well known that other mammals underwent diversification and even speciation during isolation in the glacial and periglacial conditions of the Ice Age—for example the cave bear that is accepted as a separate species, *Ursus spelaeus*, the woolly rhinoceros, *Tichorhinus antiquitatis*, and the mammoth, *Elephas primigenius*. In modern man adaptation to extreme climatic stress has resulted in the evolution of arctic and tropical desert races, which are quite distinctive. For example, the Eskimo is short and very robust; the Sudanese is generally tall and gracile. This is an instance of Allen's "rule" that states that the surface/volume ratio of different races of warm-blooded vertebrate species is greater in warm than in cold climates. The "law" holds for many human races and may well in turn account for the short but robust skeleton of the arctic-adapted West

European Neandertal man. We cannot, however, evoke this hypothesis to account for the robust "Neandertaloids" from tropical regions. However, their skeletons are not known (with the exception of the fragments from Solo and Broken Hill), and their robusticity may well have been less extreme. Whatever the explanation, and this is only one of a number that have been proposed, it is clear that during the Middle and Upper Pleistocene, *Homo sapiens* was a very variable species.

More important is the undeniable fact that, for reasons not yet understood, the more robust, heavy-jawed races either died out or were genetically swamped by the more gracile and necessarily more successful races; and that those that survived became even more gracile to produce modern man. In the case of West European Neandertal man, many groups may have become extinct, because their disappearance from the archeological record is relatively abrupt. In some sites there is no evidence of continuity with the succeeding Aurignacian people who swept in from the east during a warm period about 35,000 BP. In other sites and in other areas archeological evidence suggests cultural evolution, and the anatomical evidence suggests a less abrupt morphological transition.

But we should not see the variation of the contemporaries of Neandertal man as anything especially unusual. Among modern races, as we have said, profound variation is found in adaptation to climate. Even within tropical environments we find great differences, for example between the robust and powerful Australian aboriginal and the small and lightly built African forest pygmy (fig. 6). The racial variation during Neandertal and earlier times was probably even greater than it is today.

Many populations may have diverged in somewhat the same way as the Neandertal population as a result of some degree of isolation. The Ngandong population was possibly isolated in Java during interglacial periods when the sea level was high, while other local races may have developed specialized characters in other areas, such as Southern Africa, as they have done in the recent past. We should not be surprised to see this kind of variability at all stages of hominid evolution.

One hypothesis we should consider is that the "Neandertaloid" populations of the Middle and Upper Pleistocene provided the basis for the evolutionary differentiation of some of the major races of present-day mankind, for example, that the Ngandong population represents the ancestral stock from which the Australoid people were evolved and that Rhodesian man was the precursor of the modern Negroid races. The thesis of the polyphyletic origin of modern man, propounded from time to time by a few anthropologists in previous years, has more recently been revived by Coon in his enquiry into the origin of human races (1962). This author relies for his evidence on remains (too scanty it seems) of fossil man in China,[8] Java, Africa, and Europe, which for him suggest that the modern racial groups of Mongoloid, Australoid, Negroid (or Congoid), "Capoid," and Caucasoid peoples developed independently from the common ancestral species *H. erectus* several hundred thousand years ago. In other words, he proposes that *H. erectus* evolved into *H. sapiens* "not once but five times, as each subspecies, living in its own territory, passed a critical threshold from a more brutal to a more sapient state." Such a mode of evolution, that is, a species A becoming differentiated into subspecies A^1, A^2, A^3, A^4, and A^5, and all these evolving independently to form subspecies B^1, B^2, B^3, B^4, and B^5, that together constitute a new species B, is not a theoretical possibility because subspecies are by definition not genetically entirely independent. A species lineage evolves broadly as a unit (though different populations may be genetically isolated for limited periods in varying degrees), and the imaginary boundaries that delineate successive species *must be drawn at a point in time* (e.g., 300,000 BP for the boundary between *H. erectus* and *H. sapiens*). Coon's hypothesis could not be demonstrated even where complete isolation of the subspecies and their several descendants are known to have occurred, because their isolation did not continue: they either became extinct or were later genetically swamped. In neither case could they be said to have evolved into *Homo sapiens* at different times. If they had achieved full independence from each other and survived as well, we should now have more than one species of man, and this is not the case.

Since it is now well established that the species *H. sapiens* was in existence more than 250,000 years ago, it appears highly probable that members of the species could have migrated many times during this prolonged period over wide distances of the Old World. Thus, while it may be that, for example, representatives of *H. erectus* in China did contribute at least some genes toward the composition of the mongoloid genotype, there is no reason to suppose that they did so exclusively. It is reasonable to assume, as a matter of probabilities, that the several modern subspecies of mankind became differentiated long after *H. sapiens* had in fact become established as a definitive species.

The predecessors of the people of the Riss-Würm interglacial are known only from the few fossils of the preceding interglacial, the Mindel-Riss (Arago, Swanscombe, etc.). At this time it seems likely that though the range of human adaptations may have been less extensive, for they had certainly not penetrated so far north into the arctic regions, the species was nevertheless as diversified anatomically as it was to become later. The amount of variability we find even within Western Europe is considerable and suggests that gene flow between populations was not rapid.

The early predecessors of the African and Southeast Asian fossils are not well known. Some fragmentary remains found in China and Africa may date from this period, but they tell us little of their overall morphology and relationships. However, populations of early *H. sapiens* almost certainly existed in Southern Asia and Africa because their antecedents (*Homo erectus*), as well as their descendants, are known. This gap in the fossil record may well be filled before long.

Three

Homo erectus and *Homo habilis*

The Discovery of *Homo erectus* in Java

In the last chapter we briefly reviewed the evidence for the existence in Europe during the last interglacial period of hominids not clearly distinguishable as a separate species from *Homo sapiens*. We saw that the few skulls so far available from deposits of this date are mostly characterized by a rather heavy development of the supraorbital ridges; to this extent they certainly present a primitive appearance. We have also noted the evidence of the Swanscombe and Steinheim skulls that the species *H. sapiens* was in existence during the second interglacial period. We now have to consider a group of extinct hominids of much more primitive appearance, known from skulls, teeth, and limb bones found in the Far East, that have been for many years commonly regarded as representing a separate genus of the Hominidae, *Pithecanthropus*. Today there is a general consensus that they are more properly to be included in the genus *Homo* but specifically distinct from *H. sapiens*.

Homo erectus was for many years known only from two regions, Java (central and eastern) and China (Choukoutien, near Pekin). The Chinese fossils were at one time assigned to a distinct genus, *Sinanthropus*, but there is now general agreement that they are conspecific with the Javanese fossils. However, they present certain minor contrasts in skeletal and dental characters that may be taken to indicate at least a subspecific difference.

The first outstanding discovery of *H. erectus (Pithecanthropus)* was made by Dubois at Trinil in central Java in 1891. The remains were recovered from alluvial deposits on the bank of the Solo River, at a stratigraphic level from which has also been derived a faunal assemblage including *Stegodon trigonocephalus, Hippopotamus antiquus,* a small axis deer (*Cervus lydekkeri*), and a small antelope (*Duboisia kroesenii*). It is now generally agreed that the Trinil fauna is post-Villafranchian and that the Trinil beds are of Lower Pleistocene age. On the basis of potassium-argon datings the Trinil beds may be considered to span the period from ca. 1.3 to 0.7 million years BP.

The most important item of Dubois's original discovery was a calvaria of *H. erectus,* which, in the very flattened frontal region, the powerfully developed supraorbital ridges, the extreme platy-cephaly, and the low cranial capacity (estimated at about 850 cc), presents a remarkably primitive, apelike appearance. So much so, indeed, that some anatomists at first refused to recognize it as a hominid calvaria at all and supposed it to be the remains of a giant gibbon. In the same deposits and at the same stratigraphic level (but 15 meters upstream from the calvaria) Dubois found a complete femur that, in its size and general confirmation, is very similar to the femur of *H. sapiens.* Some doubt was naturally expressed (mainly because of their apparent incongruity) whether the femur and calvaria belonged to the same individual and whether the femur was really indigenous to the Trinil deposits. However, the accumulation of evidence speaks so strongly for their natural association that this has become generally accepted. In the first place, there seems little doubt that the femur was actually found in situ in the Trinil beds (and, according to Hooijer [1951], the field notes kept by Dubois show that his excavations were carried out systematically and accurately). Second, it has been reported by Bergman and Karsten (1952) that the fluorine content of the femur is equivalent to that of the calvaria and also of other representatives of the Trinil fauna, an observation that is in conformity with the geologic evidence that they were contemporaneous. Third, remains of four other femora of similar type in the Leiden Museum have been found among fossils collected from Trinil deposits (three of them were described by Dubois 1932), and these also show a fluorine content compatible with a Middle Pleistocene date. Thus they confirm that at this early time there existed in Java hominids with a type of femur indistinguishable from that of *H. sapiens*, though all the cranial remains so far found emphasize the extraordinarily primitive characters of the skull and dentition. Finally, portions of seven femora of the pithecanthropines from Choukoutien, though they carry minor differences from the Trinil femora (see below), show that in the Chinese representatives of this population a femur of an almost modern human type was also associated with the same primitive features of the skull and dentition. The combination in *H. erectus* of limb bones of modern type with cranial and dental characters of a primitive type is worth emphasizing, for it illustrates an important principle of vertebrate evolution—that the progressive modification of the several somatic functional complexes may (and frequently does) proceed at different rates. This principle is well recognized by paleontologists but appears occasionally to have misled anthropologists. Such differential rates of somatic evolution may lead to structural contrasts that give an appearance of incongruity, and they are liable to be regarded as "disharmonies" because they do not conform with the sort of correlations that studies confined to living species may lead one unconsciously to expect. The true affinities of such fossil forms may thus be overlooked and misinterpreted simply because they do not exhibit the particular combination or assemblage of characters of which we have created a rather rigid mental image from our preoccupation with the comparative anatomy of living forms (the latter, in many cases, being only the relics of a much greater diversity in the past). Differential rates of somatic evolution obviously need to be taken into account in deciding the taxonomic relevance of characters that are selected for assessing the phylogenetic status of fossil types (see p. 29).

For many years after Dubois's original discoveries in Java more than half a century ago, no further remains of *H. erectus* (*Pithecanthropus*) were found, in spite of the efforts of the Selenka expedition of 1907-8. Then, during the few years preceding World War II, further important discoveries were made by von Koenigswald (1936, 1937, 1938, 1940). With the exception of an immature skull found at Modjokerto in eastern Java, these new specimens come from Sangiran in central Java, about 50 km from Trinil. They were derived from two horizons, one corresponding to the Trinil deposits and consisting of sandstones and conglomerates, the other an underlying stratum of black clay termed the "Djetis layer." The latter contains a faunal assemblage definitely older than that of the Trinil layer, characterized by *Epimachairodus zwierzyckii, Leptobos cosijni, Nestoritherium, Megacyon*, and *Cryptomastodon*; and by some authorities this has been taken to indicate an Early Pleistocene date. Preliminary potassium-argon determinations suggest that an age range of 1.3-2.0 million years BP would be reasonable for the Djetis strata. Hooijer (1951) has pointed out that there is some doubt regarding the provenance of certain of the *H. erectus* remains discovered by von Koenigswald, that is, whether their stratigraphic position is referable to the Trinil or Djetis horizon, because not all were found in situ. However, it seems reasonably certain that at least some of them came from the Djetis beds and therefore that the antiquity of *H. erectus* in Java may have extended back to the Early Pleistocene.

After World War II, further discoveries were made by Marks, Sartono, and Jacob (see Oakley et al. 1971-77) so there is now a relatively large sample of human fossils from each of the two main levels of Lower Pleistocene age. The discoveries are summarized in table 1. (See also Jacob 1973, 1975.)

The whole problem of *H. erectus* in Java has unfortunately been much confused by the multiplicity of the taxonomic terms that have been applied to the various remains. The immature skull found at Modjokerto in 1936 was originally named *H. modjokertensis*, though there was no reason to suppose that it was other than that of a young individual of *H. erectus*. Von

Table 1 Fossil Hominids from Java

Stratigraphic Unit	Sites	Age Bracket
Trinil (Kabuh Formation)	*Sangiran*	1.3-0.7 m.y. BP
	S2 Adult female calotte (1937)	[K-Ar date c.
	S3 Juvenile calotte (1938)	0.83 m.y.]
	S8 Right mandible (1952)	
	S10 Adult male calotte (1963)	
	S12 Old male calotte (1965)	
	S15 Maxilla (1969)	
	S17 Cranium (1969)	
	S21 Mandible (1973)	
	Trinil	
	T2 Calotte (1892) [=*Pithecanthropus*]	
	T3, T6-9 Femora	
	Kedung Brubus	
	KB1 Right juvenile mandible (1890)	
Djetis (Putjangan Formation)	*Sangiran*	2.0-1.3 m.y. BP
	S1a Right maxilla (1936)	[K-Ar date c.
	S1b Right mandible (1936)	1.9 m.y.]
	S4 Adult male calvaria and maxilla (1938-39) [=*P. robustus*]	
	S5 Right mandible (1939) [=*P. dubius*]	
	S6 Right mandible (1941) [=*Meganthropus*]	
	S9 Right mandible (1960)	
	S22 Maxilla, mandible (1974)	
	Modjokerto	
	M1 Child, 7 years, calvaria (1936)	

Koenigswald was also led to apply the specific term *modjokertensis* to other remains of *H. erectus* found in the Djetis layers at Sangiran "for morphological and stratigraphical reasons." However, he did not define any morphological differences that might justify a formal diagnosis of this new species, and by themselves the stratigraphic reasons are not relevant. For though the stratigraphic evidence certainly demonstrates that *Homo erectus*

existed in Java during long periods of the Lower Pleistocene it
does not preclude the possibility (or even the probability) that
during this time it was represented by a single species, *H. erectus*.
It may be noted that one fossil find (S4) was given the specific
name *robustus* by Weidenreich, again with no morphological
justification: only the back part of the skull and the maxilla of
this specimen were found. The back part of the skull shows no
distinctive features whereby it can be separated specifically from
H. erectus, and there was no other material available with which
the maxilla could be compared. Of the two mandibular fragments
found at Sangiran in 1939 and 1941, the latter has been referred,
because of its large size, to another genus altogether, *Meganthropus palaeojavanicus*. The former (S5), though very similar in
its general size and proportions, was given a specific distinction
partly because of the wrinkled character of the enamel on the
molar teeth (which may, however, be no more than the expression
of an individual variation) and partly, it seems, because the
specimen is rather poorly preserved, so that certain details of the
dental morphology are indeterminate. As regards the 1941 mandible (S6), the generic separation of *"Meganthropus"* from *H.
erectus* (*Pithecanthropus*) can hardly be justified on the basis of
dental morphology; the two premolars and the first molar that
have been preserved in this specimen conform in their total
morphological pattern to a hominid dentition of the type found in
the 1936 mandibular fragment. The large size of the teeth and the
preserved portion of the body of the mandible is certainly
striking; but, compared with the 1936 mandible, it does not
indicate a range of variation exceeding that found (for example)
in the single species *H. sapiens* (fig. 11). It may readily be admitted that the available fossil material from Java is not yet
adequate to decide finally whether there was more than one
genus, or more than one species, of hominid living in Java during
the Early Pleistocene. But from general considerations the probabilities seem to be against such a conclusion, and (it must be
emphasized again) in spite of a considerable range in the degree
of robusticity, there is at present no really convincing morphological basis for the recognition of more than one species, *H.*

erectus. To avoid further confusion, we shall here refer to all the specimens simply by the specific term *H. erectus,* while recognizing that there may well have been more than one subspecies or race.[1]

The Morphological Characters of the Javanese Representatives of *Homo erectus*

It has already been mentioned that the calvaria of *Homo erectus* in Java is known from seven adult specimens and one immature skull. All the adult specimens are sufficiently complete to permit fairly close estimations of the cranial capacity by making endocranial casts and restoring the missing parts in proportion. From the Trinil level, there are six skulls from which an estimate of cranial capacity can be calculated. Tobias, who has reviewed and summarized the data (1971) records a range of 775–1,029 cc and the data give a mean capacity of 906 cc. The endocranial capacity of only one adult skull (S4) can be measured from the lower Djetis

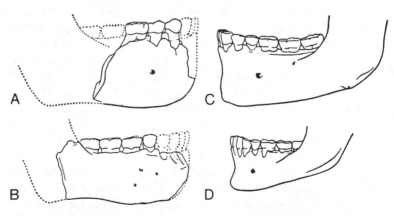

Fig. 11. The mandibles (reconstructed) of "*Meganthropus palaeojavanicus*" (A) and *H. erectus (B)*, compared with two mandibles of *Homo sapiens* (an Australian aboriginal [C] and a European [D]) to indicate that the range of variation in the size of the Javanese fossils does not exceed that of a single species of *Homo.* Approximately one-half natural size.

level, and the figure obtained is 750 cc. The cranial capacity of
the Modjokerto child skull from the Djetis level is estimated to be
no more than 650 cc; and, by reference to data concerning the
brain growth of modern man and anthropoid apes, it may be
inferred that, given an age for the child of six years, brain growth
would be nearly complete and would be unlikely to have developed beyond about 750 cc in the adult. This would make a mean
value for the small Djetis sample of 750 cc. These figures may be
compared with a range for *Homo sapiens* of ca. 1,000-2,200 cc
and a mean of 1,350 cc.

As might be expected, a number of authorities have examined
in detail the surface features of the endocranial casts of *H. erectus*
in an attempt to find therefrom some clue to the intellectual
status of these primitive hominids. Let it be said at once,
however, that these studies have had very disappointing results,
for the information to be obtained from endocranial casts
regarding brain functions is strictly limited. There are two
main membranes lining the brain and interior of the skull
(arachnoid and dura mater respectively), separated by fluid
that acts as a cushion to protect the brain from impact. As a
result, impressions of the brain itself are hardly identifiable on
the inner table of the cranial bones.[2] The most striking impressions are those of the meningeal arteries and venous sinuses, as
well as the main subdivisions of the brain into cerebrum,
cerebellum, and, to some extent, the brain stem. Some components of the cerebrum also leave their mark, and the temporal
lobe is clearly defined by the sylvian fissure (see fig. 16). Some of
the convolutions of the cortex are faintly reflected in the endocranial wall, but it is not usually possible to identify them with
any certainty. In some specimens, the simian sulcus or parieto-
occipital fissure, which separates the parietal and occipital lobes,
can also be identified.

In the past, many inferences have been made regarding the
acquisition of articulate speech, the degree of manual skill, the
ability to learn from experience, and other mental faculties in
fossil hominids; these must now be discounted. This applies
equally to the assumption that right- or left-handedness can be
inferred from a consideration of the asymmetry of the cerebral

hemispheres (Clark 1934). In hominids and the large anthropoid apes the sulci usually do not impress themselves clearly on the endocranial aspect of the skull except near the frontal and occipital poles of the brain and in the lower temporal region (for a useful critique on the interpretation of endocranial casts, see Hirschler 1942). All that can be said of the brain of *H. erectus* is that the low average size is presumably related to a rather low level of general intelligence (cf. fig. 6, p. 119). By relating cranial capacity to femur size, Brummelkamp (1940) has tried to assess the degree of "cephalization" or relative cranial capacity of fossil hominids. While the cephalization stage of less ancient fossil specimens of *Homo* (e.g., those found at La Chapelle-aux-Saints, Combe Capelle, Spy, Grimaldi, and Chancelade) is found not to differ from that of Recent *H. sapiens*, in the case of *H. erectus* (from both Java and China) it is lower by a factor of $1/\sqrt{2}$.

The main characteristics of the skull of the Javanese representatives of *H. erectus* may be stated briefly as follows (figs. 12 and 22). The calvaria shows a very marked degree of platycephaly, with the maximum breadth low down in the temporal region. From this level the lateral walls of the cranium slope upward and medially to the position of the parietal eminence and then medially with a slight upward inclination to a median sagittal ridge. The supraorbital torus is developed to an exaggerated degree compared with modern or even Neandertal man and is bounded behind by a receding frontal contour and a very marked postorbital constriction. The latter is associated with an inward curvature of the anterior part of the temporal squama, which accentuates the apelike appearance of the skull as a whole. The occipital torus is massive (particularly in S4), and in all cases projects backward well beyond the level of the supraoccipital squama. The nuchal area of the occipital bone is relatively extensive (evidently for the attachment of a very powerful nuchal musculature) and slopes backward and upward to the occipital torus. The tympanic region and mandibular fossa are hominid in their general characters, but the rather weak development of the articular eminence in S4 and the rounded contour of the external

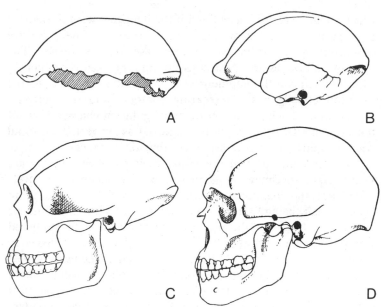

Fig. 12. Comparison of the lateral views of the *Homo erectus* skulls found in Java with that of an Australian aboriginal. *A, Homo erectus* (T2); *B, H. erectus* (S2); *C,* reconstruction of *H. erectus* (S4) (modified from Weidenreich); *D,* Australian aboriginal. Approximately one-quarter natural size.

auditory meatus in S2 may be regarded as primitive characters of a somewhat simian type. The mastoid process is variably developed, and in S4, where it is large, the apex is strongly deflected medially. The petrous bone, as seen from the endocranial aspect, is rather massive, and the cranial wall is everywhere of unusual thickness.[3] The foramen magnum (as far as can be judged from the crushed skull base in S4) is situated as far forward in relation to the total cranial length as it is in *H. sapiens*.

The infant skull from Modjokerto shows features of particular interest. Though the cranial walls are very thin (in the parietal region up to 3 mm, and elsewhere even less), the degree of ossification of the tympanic region suggests an age corresponding to five or six in *H. sapiens*. But the precise age of this infant, which could be better determined if the dentition were present, is perhaps not a matter of great importance, for the primitive characters of the skull are in any case very obtrusive. The supraorbital ridges are already assuming a marked prominence, with the incipient development of a postorbital constriction; the forehead already has a retreating contour; and the occipital region shows the development of the angulation characteristic of the adult skull. If the individual is as old as six years, the small cranial capacity of 650 cc assumes great significance.

The palate and part of the facial skeleton of *H. erectus* are known from specimen S4. The size of the palate is relatively enormous, the maximun maxilloalveolar width, according to Weidenreich, being 94 mm. The maximum width, it should be noted, is at the level of the last molar teeth and may be compared with the maximum width, 80 mm, of the palate of the Rhodesian skull, which is at the level of the second molars. The facial aspect of this specimen shows a pronounced alveolar prognathism, a great breadth of the anterior nares (which is near the uppermost limit so far recorded for *H. sapiens*), and an extensive maxillary sinus. The lower margin of the nasal aperture is bounded by a simple margin, with no sulcus or fossa prenasalis. The 1936 Sangiran mandible (S1b) is heavily constructed, with a very thick and sloping symphyseal region, and there is no indication of a mental eminence. It shows three mental foramina—a very unusual feature in *Homo sapiens* but quite common in anthropoid apes. In its general build the mandible conforms quite well with the maxilla S4. The mandibular fragments found at Sangiran in 1939 and 1941 (S5, S6) provide evidence of a lower jaw even more massive; but (as already noted) there seems no sound morphological reason for not including them in the species *H. erectus*. In both specimens the mental foramen is single and is situated about midway between the alveolar and lower margins of the mandible —a hominid feature that contrasts rather strongly with the low position of the foramen in the large anthropoid apes. In the 1941 mandibular fragment (S6) the symphyseal region is exceedingly thick, but in its general contour in sagittal section it resembles the Mauer mandible (described below) quite closely, and in the high position of the foramen spinosum it also shows a hominid, rather than a pongid, feature. In none of these mandibular specimens found in Java is there any indication of a "simian shelf."[4]

That in its total morphological pattern the dentition of the *H. erectus* group of fossils from Java is characteristically hominid in type has not been a matter for dispute. The contours of the dental arcade, the shape, size, and morphological details of the canines and premolars, and the morphological details and mode of wear of the molars, taken together, provide a marked contrast with the pongid type of dentition. On the other hand, the specimens available show certain primitive features, in which some approach is made to a pongid level of evolutionary development. For example, in the specimen S4, the upper canine, though not very large relative to the adjacent teeth, and spatulate rather than conical in general shape, projects well beyond the level of the premolars and shows a well-marked attrition facet on its anterior margin, which indicates that it overlapped the lower canine to a slight degree. Moreover, it is separated from the socket for the lateral incisor by a distinct diastemic interval. Von Koenigswald noted that, of the two other maxillary fragments from the Djetis layers of Sangiran, one shows evidence of a similar diastema, while in the other there is a contact facet on the canine that makes it clear that there was no diastema in this specimen. The primitive trait of a diastema in the upper dentition was thus not a consistent feature of the Javanese representatives of *H. erectus*. From isolated teeth (not hitherto described in detail), von Koenigswald (1950) concluded that the central upper incisors are "extremely shovel-shaped, by far surpassing the condition observed in *Sinanthropus*." The upper molars, premolars, and canine form converging rows on either side of the palate, with only a slight degree of curvature. The first upper premolars are provided with three roots, as is normally the case in the Pongidae but uncommon in *H. sapiens*. The upper molars are large, the second molar being larger than the first.[5]

Of the lower dentition, the mandibular fragment S1b shows in a fairly well-preserved state the second premolar and the three molar teeth. The latter are noteworthy for their large size and for the fact that their length increases progressively from front to back. This is a primitive feature very rarely found in *H. sapiens*. The alveolar socket for the first premolar is simple and indicates the single root construction characteristic of the Hominidae. The socket for the canine tooth is very small (particularly when considered in relation to the massiveness of the jaw) and is placed medially to the anterior margin of the first premolar socket. Von Koenigswald has commented on the weak development of the incisors and canines when compared with the large size of the postcanine teeth, a character that is even more marked in the australopithecine material from South Africa (see p. 151). No isolated lower canines or incisors from Java have so far been described in detail. In the large mandibular fragment from Sangiran ("*Meganthropus*" S6), the two premolars and the first molar are well preserved but are somewhat worn. The first premolar is of the hominid type, bicuspid with well-marked anterior and posterior foveae, and shows no tendency toward the sectorialization that is in general characteristic of Pongidae (fossil and Recent). There is a surface indication of a subdivision of the root, but no separation of the component elements. The socket for the canine shows that this tooth was surprisingly small, the total length of the root being certainly not more than 20 mm. Von Koenigswald (1950) mentioned an isolated lower canine from Java, which he also referred to "*Meganthropus*," but gave no details of this specimen except to say that it is small (presumably relative to the other teeth) and "in no way different from the canine of modern man except for the size." It should be noted that the preserved teeth in the large 1941 Sangiran mandible are very similar to those of the *H. erectus* S1b mandible, except for their overall dimensions. Weidenreich has also emphasized their strong resemblance to the corresponding teeth of the Pekin representatives of *H. erectus*, particularly in the arrangement of the cusps, the distinctness of the anterior and posterior foveae, and the development of the cingulum of the first premolar tooth. Some authorities have interpreted the large mandibular fragments from Sangiran as evidence for the existence in Java during the Pleistocene of "giant" hominids. This seems to be a misapplication of the term "giant," which is commonly taken to refer to stature. But a large hominid jaw does not imply a giant individual. On the contrary, so far as other paleontological evidence goes, there is some reason for assuming a negative correlation between the size of the mandible and the total stature. Certainly, in the case of the Javanese fossils, the femora, which have been assigned to the Trinil horizon, provide no evidence of great height —estimates based on these specimens indicate a stature of about 5 feet 8 inches or less, and in no case more than 5 feet 10 inches. It has already been noted that all these thigh bones (including the complete femur originally found by Dubois in 1891 and the incomplete shafts later discovered among other fossil material from Java) are similar to those of *H. sapiens*.

The Morphological Characters of the Chinese Representatives of *Homo erectus*

It was entirely due to the care and foresight of Professor von Koenigswald that the important discoveries of *H. erectus* that he made in Java shortly before the last war were preserved intact. On the eve of the Japanese invasion of Java, he distributed the valuable specimens among his friends of the local population, and when he himself was liberated from captivity with the final defeat of the Japanese, he was able to retrieve them intact. In the case of the remains of *H. erectus* found at Choukoutien near Pekin, the story was, unhappily, very different. At the end of the war no trace of the fossils could be found. It is generally supposed that they had been crated for dispatch to a safe area and that the ship on which they had been loaded was sunk in the early stages of the war. Fortunately, however, detailed and comprehensive descriptions had been published by Davidson Black (1930) and Weidenreich (1936, 1937, 1941, 1943), richly and accurately illustrated by drawings, photographs, and radiographs, and casts are available for study.

There is no need to recount in detail the history of the first discovery of *H. erectus* (*Sinanthropus*) in Pekin, for the story has been well told elsewhere (see Elliot Smith 1931 and Keith 1931). A single lower molar tooth found in 1927 at Choukoutien led Davidson Black—at that time professor of anatomy at Pekin University Medical College—to create a new genus and species of the Hominidae, *Sinanthropus pekinensis*. This decision, based as it was on one tooth only, was rightly met with some skepticism; but Black's inference that the tooth provided evidence for the existence in China during the Pleistocene of a primitive type of hominid showing certain pongid characters displayed in a remarkable way his perspicacity as a comparative anatomist. For two years later, in 1929, an uncrushed and almost complete calvaria of a very primitive type was found during excavations at the same site. Its similarity to *H. erectus* of Java (T2) was very obvious from the first, and subsequent studies and further discoveries finally made it clear that the Javanese and Chinese fossils were not generically, or even specifically, distinct (von Koenigswald and Weidenreich 1939). The generic term *Sinanthropus* has therefore now been discarded. There are certain minor differences, however, and, as already mentioned, these are held to justify a subspecific distinction.

Chinese botanists' pollen analysis of the cave deposits at Choukoutien, which contained this remarkable collection of fossil human remains, shows clearly that they were deposited during a long warm spell of the Pleistocene. The faunal remains suggest a Middle Pleistocene age, but no radiometric dates have been determined. The general consensus equates this warm spell with either the European Mindel-Riss interglacial or the preceding warm phase (the Inter-Mindel). This means the deposits date from somewhere between 500,000 and 300,000 years BP and are certainly much more recent than all the remains of Java man.

The Chinese variety of *H. erectus* is known from fifteen calvariae (or fragments of calvariae), as well as portions of facial bones, many teeth, and a few limb bones, representing some fifty individuals. The main features of the skull are as follows (fig. 13): the cranial capacity (based on Weidenreich's estimates from five calvariae) ranges from 915 to 1,225 cc, with a mean of 1,043 cc. So far as this limited material goes, therefore, it appears to

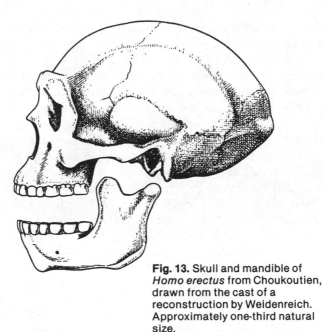

Fig. 13. Skull and mandible of *Homo erectus* from Choukoutien, drawn from the cast of a reconstruction by Weidenreich. Approximately one-third natural size.

indicate that the mean cranial capacity may have been about 150 cc greater than that of the Java Trinil specimens, and at its upper levels the brain volume actually comes well within the normal range of variation of *H. sapiens*. All the calvarial specimens show a marked platycephaly, with heavily constructed supraorbital and occipital tori and a rather well-marked sagittal ridge. They are relatively homogeneous in their general shape but differ considerably in size (possibly an expression of sexual dimorphism). Viewed from behind, they are broader at the base, the width of the skull diminishing from the level of the interauricular plane upward. Compared with *H. erectus* of Java, the platycephaly is a little less extreme (though it should be emphasized that the cranial height can be only approximately estimated in the Javanese fossils), and there is a distinct, though slight, convexity of the frontal squama. The frontal sinus is unusually small in some of the specimens (in the Javanese types it is very large), though the

other accessory air sinuses, for example, in the maxilla and the
mastoid process, are well developed. The bones of the skull wall
are of massive thickness, the latter (according to Weidenreich)
being due mainly to a thickening of the outer and inner tables. All
the bones of the facial skeleton are likewise heavily constructed.
The mandibular fossa is unusually deep, bounded in front by a
conspicuous articular eminence. The tympanic plate is very thick
and is disposed more horizontally than in *H. sapiens* (resembling
in this respect *H. erectus* of Java, as well as the large anthropoid
apes). The auditory aperture is wide; it varies in its shape but
commonly has a transversely elliptical form. The mastoid process
is relatively small. The nasal bones are said to exceed in width
those of *H. sapiens*, and the nasal aperture (which is not bounded
by a prenasal groove) is conspicuously broad. The palate shows
the typical hominid contour and is not exceptionally large in its
surface area. The cranial sutures appear to close earlier than they
do in modern man.

The mandible is robust, with a bicondylar width that reaches
the upper limit of that of modern races such as the Eskimo (and
perhaps exceeds it). The angle of inclination of the symphyseal
axis, according to Weidenreich, is about 60° and in this character
corresponds with the Mauer mandible (see below, p. 115). There
is no mental eminence. The digastric fossae (for the attachment
of the digastric muscles to the back of the symphyseal region of
the jaw) are elongated and narrow and do not extend to the
vertical inner surface of the mandible as in *H. sapiens*. The
mental foramen is multiple in all the specimens found; indeed,
there may be as many as five foramina—a remarkable feature.
The ascending ramus is broad, and the muscular markings for
the masseter and pterygoid muscles are strongly developed. The
articular surface of the condyloid process shows no feature that
differentiates it from that of other species of *Homo*. The coronoid
process is broad and thick and presumably served to attach a
rather powerful temporal muscle.

The endocranial cast has been studied by Davidson Black
(1933) and also by Shellshear and Elliot Smith (1934). From the

accounts of these authors it is very closely similar to that of the Javanese specimens of *H. erectus* and shows no distinguishing features.

The dentition has been studied in the greatest detail by Weidenreich (1937) on the basis of 147 teeth found at Choukoutien, representing probably thirty-two individuals. Of these specimens, thirteen belong to the deciduous dentition. The teeth are large by modern standards, the enamel surface is frequently complicated by rather elaborate wrinkling, and the basal cingulum (the thickening of enamel at the base of the crown of the tooth) is particularly well developed. The canines show considerable variation in size, shape, and robustness, but those of the upper dentition are frequently large and conical. However, it is important to recognize that, in spite of these primitive features, there is no evidence that they projected to any marked degree beyond the level of the adjacent teeth,[6] and in the early stages of attrition they quickly became worn down to an even, flat surface, in conformity with the occlusal plane of the dentition as a whole. Thus the upper and lower canines did not overlap, as appears to have been the case in some of the Javanese representatives of *H. erectus*. The first upper premolar may have two roots, with a surface indication of a third. The first lower premolar is a nonsectorial tooth of bicuspid shape, with a strong development of the lingual cusp. The two cusps, in fact, merit the term "subequal"; and herein the tooth contrasts strongly not only with that of the anthropoid apes, which carries only one cusp, but also with modern *H. sapiens*. For in the latter the lingual cusp seems to have secondarily undergone a considerable retrogression, being usually much smaller than the buccal cusp. The root is single but shows evidence of the fusion of two separate roots. The molar teeth show no special features, apart from their size and the strong development of their main cusps and some secondary cuspules. They show a flat wear with even a moderate degree of attrition; in this feature they contrast definitely with the anthropoid apes. The second lower molar is slightly larger than the first, and the third molar is usually the smallest of the three. The

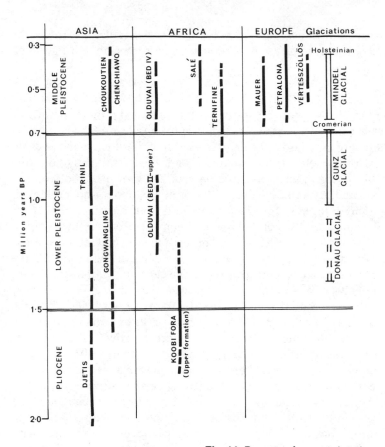

Fig. 14. Ranges of uncertainty in the ages of *Homo erectus* fossils.

dental arcade shows the even curvature characteristic of *Homo*, and in none of the specimens is there a diastema. There is some evidence from immature jaws that in the eruption of the permanent dentition the second molar appeared before the canine, a sequence characteristic of the anthropoid apes, but also of certain modern races of man. The deciduous dentition is distinguished by the fact that the first milk molar is rather strongly compressed from side to side and by the sharply pointed shape of the canine, which also has a well-developed cingulum. In both these features the deciduous dentition may be said to show a slight degree of morphological approximation to that of the Pongidae.

A well-preserved mandible of about the same date was excavated by Chinese workers at Chenchiawo, Shensi Province, in 1963. Belonging to an aged female, it conforms in most of its characters to the Choukoutien specimens, but is remarkable in having congenitally absent third molars on both sides (Aigner and Laughlin 1973). As is well known, this feature is commonly found in some races of modern man.

The limb bones found at Choukoutien comprise portions of seven femora, two humeri, a clavicle, and one of the carpal bones (os lunatum). Of the seven femoral remains found at Choukoutien, one consists of almost the entire diaphysis, one of the proximal half of the diaphysis, and the others are still more fragmentary (Weidenreich 1941). They show a marked degree of platymeria (though not more so than may occur in *H. sapiens*), and the wall of the shaft is unusually thick, the medulla narrowed. Weidenreich expressed the opinion that in certain features of the curvature of the shaft the bones differ from those of modern man; however, it cannot be held to constitute by itself a very significant difference. At the same time, the femoral shafts certainly lack the robustness and curvature characteristic of Western Neandertal man. From the most complete shaft (femur IV), the total length of this bone has been estimated by Weidenreich to have been 407 mm; and from this the total standing height of the individual is computed to have been 5 feet 1½ inches. If this were to be taken as representative of the local population as a whole (obviously an unsafe assumption), it would suggest that the Chinese population of *H. erectus* was shorter in stature than the Javanese population of the same species. The humeral fragments found at Choukoutien consist of a diaphysis of one bone and a small portion of the distal end of another one. Apart from the thickness of the shaft walls, they show no distinguishing features at all. The clavicle and the os lunatum are also very similar to those of *H. sapiens*.

An important skullcap with maxilla was found in China in 1964 at Gongwangling, Shensi Province. Though incomplete and weathered, the skull clearly is remarkably similar in many features to the smaller-brained robust material from the Djetis levels in Java, though it is said to be even more primitive. It has very thick vault bones and a very well developed supraorbital torus, giving an impression of immense robusticity. The endocranial volume is estimated at 780 cc. On faunal grounds, the site precedes the Choukoutien deposits by perhaps 400,000 years, which suggests an age in the region of 0.7 million years BP (Aigner and Laughlin 1973).

So far as the scanty fossil material permits a comparison of the Javanese and Chinese representatives of *H. erectus*, it is probably true to say that the former were more primitive in their smaller cranial capacity, more marked platycephaly, greater flattening of the frontal region of the skull, more heavily constructed mandible, less pronounced curvature of the dental arcade, larger palate, a tendency to slight overlapping of the canines with the occasional presence of a diastema in the upper dentition, and the relative length of the last lower molar. It must, however, be admitted that though the Choukoutien femoral fragments show certain distinctive features (which they share with the Olduvai femur, see below) these features are not present in the Trinil femora, which are fully modern in form. This inconsistency, pointed out by Day and Molleson (1973) throws some doubt on the antiquity of the Trinil femora and their supposed association with the famous calotte (p. 91).

It is also clear that in Java *H. erectus* extended back to a greater antiquity than in China, for while (as already noted) the faunal correlations of the Djetis deposits have been taken to indicate an Early Pleistocene horizon, the deposits at the Gongwangling site are closer to Trinil in age, and the Choukoutien deposits are certainly no earlier than the Middle Pleistocene. Thus there are three well-separated chronological groupings: the Djetis specimens (ca. 2.0–1.8 million years BP), the Trinil specimens (ca. 1.0–0.7 million years BP), and the Choukoutien specimens (ca. 0.5–0.3 million years BP). On purely morphological criteria, however, the three groups appear closely related, the distinction being certainly no more than a subspecific one.

It is not known certainly what degree of cultural development had been attained by *H. erectus* in Java, for the evidence is entirely negative. No implements were found in direct association with the skeletal remains. However, chopping tools, hand axes, and primitive flake tools of the Patjitanian industry have been recovered from deposits of a slightly later age, and it is not improbable that such tools were actually made and used during the Lower Pleistocene by hominids of the *H. erectus* type in Java. At Choukoutien, in the same deposits yielding remains of *H. erectus*, crude cores and trimmed flakes of quartz and silicified rocks were found, forming a local industry of a simple, but fairly uniform, character. Animal bones were also found, broken and chipped, apparently by design for use as tools. Finally, the remains of hearths throughout the deposits, as well as charred animal bones, provide evidence that these early hominids were familiar with the art of making and using fire for domestic purposes, while the nature of their diet is indicated by collections of deer bones and hackberry seeds. It appears, therefore, that, in spite of the crudity of their stone and bone industry, the Middle Pleistocene population of *H. erectus* in China had already developed a very active communal life; and it is of particular interest that at this early time they had already acquired the intelligence and skill to use fire for cooking. Indeed, many of the human bones were found closely associated with animal food remains in and near the hearths, and this possibly constitutes evidence of cannibalism. The base of each skull is broken, which may have been done to extract the brain. The relative absence of human postcranial remains suggests that the cave dwellers discarded the remainder of the body outside the cave. Ritual cannibalism of this kind has been recorded among living tribes of Southeast Asia and elsewhere. There is little reason to suppose that Pekin man was not culturally advanced, and he certainly demonstrates many behavioral traits characteristic of modern man.

Homo erectus in Africa and Europe

For fifty years or more it had been tacitly assumed that the primitive type of man now called *Homo erectus* was confined in geographical distribution to the Far East. More recently, evidence has accrued to make it reasonably certain that the same type also existed at about the same time in Africa during the Lower and Middle Pleistocene, and probably in Europe also.

In 1954 there were discovered by Arambourg (1957) at Ternifine in Algeria three mandibles and a parietal bone in deposits dating from early Middle Pleistocene times. The jaws are remarkably robust and lack a chin eminence, the molar teeth are very similar to those of *H. erectus* of Pekin, the canine teeth (as inferred from their sockets) were exceptionally strong, and the parietal bone conforms in size and curvature with known specimens of *H. erectus*. Indeed, there can be little doubt, so far as morphological evidence goes, that these remains do belong to the same species. It is unfortunate, therefore, that the group they represent was at first designated *Atlanthropus mauritanicus*, with the implication that it was not only specifically but even generically distinct. This is but another example of the terminological confusion in paleoanthropology resulting from the misapplication of taxonomic principles.

A much more recent discovery from the same region is that of a calvaria from Salé, Morocco, found in 1971. It was associated with a maxillary fragment and a natural endocranial cast. The age of the skull is not yet established, but it is almost certainly more recent than the Ternifine remains and is probably of Late Mindel age (Jaeger 1975).

The frontal bone is broken just behind the supraorbital torus, which is missing, but the bone carries the marked postorbital constriction and sagittal ridge typical of *Homo erectus*. The vault is long and low, but the occipital torus is not well developed and the occipital angle is open, so that the inion lies *below* the opisthocranium. This is far closer to the Vértesszöllös pattern than to the Choukoutien pattern (see below). However, the skull has a cranial capacity of only ca. 945 cc and the mastoid processes are very small. The dentition of the left maxilla is very worn but again appears closer to the European sample (to be described) than to the Asian members of the species in size and form. There is a well-developed cingulum on the labial side of the second and third upper molars.

Though it is characterized by many important features of *H. erectus*, this skull clearly carries some more modern features as well, which ally it with the early *H. sapiens* population of Europe. The relationships of this unusual skull will be discussed later in this chapter.

Farther south in Africa, at Olduvai Gorge in Tanganyika, Louis Leakey (1961) discovered, in 1960, a calvaria from a stratigraphic horizon containing Acheulian tools in upper bed II, and with an antiquity of perhaps 1.0 million years. Thus it appears to have been more or less contemporaneous with the Trinil Javanese representatives of *H. erectus*. So far as can be inferred from a brief inspection of this important specimen and from published photographs (for it has yet to be described in full detail), the calvaria is so similar to some of the skulls of *H. erectus* from China that it would be difficult to justify a specific distinction. The cranial capacity is 1,067 cc.

From a somewhat higher level in the gorge, bed IV, a femur and innominate bone were recovered in 1970. These deposits are believed to date from about 0.5 million years BP and the femur is reported to be extremely similar to the Choukoutien femoral fragments (Day 1971): it has marked platymeria, a distal point of minimum breadth, and some convexity of the medial border of the shaft. The innominate associated with it (the first of *Homo erectus* to be discovered) is very robust, with a large acetabulum and massive vertical iliac pillar, unmatched in any known modern innominate. It is incomplete, but it clearly belonged to a powerful individual of the genus *Homo* and is quite distinct from the much earlier pelvic remains of *Australopithecus* (see chap. 4).

Since 1967, Richard Leakey and G. L. Isaac have led expeditions to collect fossil hominid remains from many sites in an area called Koobi Fora, in Northern Kenya, east of Lake Turkana (formerly Lake Rudolf). Their undertaking has resulted in more discoveries of fossil hominids than any previous expedition of this kind. Their numerous finds can be divided, for convenience, into two genera: *Australopithecus* and *Homo* (Leakey and Leakey 1978). For the present we shall briefly consider specimens clearly belonging to the latter genus and derived from the upper member of the Koobi Fora Formation.

Among this small group, found during 1974–75, is an exceptionally well preserved skull (ER 3733): excluding Petralona (see below), it is the finest cranium of *Homo erectus* known (Leakey 1976). Radiometric dates from tuffs in the area indicate a Lower Pleistocene age of between 1.2 and 1.6 million years BP. The cranium is large, with large supraorbital tori and little postorbital constriction. There is a marked postglabellar sulcus, and the frontal rises steeply from behind it to reach the vertex at the bregma (fig. 15). The skull decreases in height from bregma, and the occipital bears a pronounced torus where the occipital and nuchal planes are sharply angled. The greatest breadth is low, and the temporal lines are strongly marked, the vault bones very thick, the temporal fossae small. The face is partly preserved with deep and wide zygomatic bones; the palate is rather rectangular, with a high roof. The dentition is badly damaged and has not yet been described. In summary, the cranium is reported to be strikingly like those from Choukoutien, though its age is apparently much greater. The endocranial capacity is estimated to be 800 cc.

A second fossil from approximately the same level is also of great interest. It is an innominate from a large, robust male (Leakey 1976). The pubic and ischial rami are missing, but most of the ilium is present. It shows striking similarities to the Olduvai innominate described above, though it is almost certainly a good deal older. It has many features in common with modern man, but it also carries the heavy vertical iliac pillar and large acetabulum of the Olduvai specimen. At the same time the lateral flare of the ilium is reminiscent of *Australopithecus*. Though this bone and the Olduvai ilium are not directly associated with skulls attributable to *Homo erectus*, it does seem reasonable to suppose

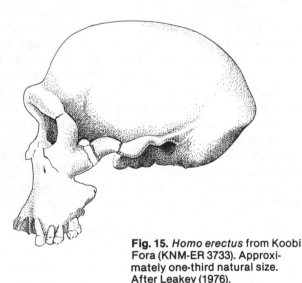

Fig. 15. *Homo erectus* from Koobi Fora (KNM-ER 3733). Approximately one-third natural size. After Leakey (1976).

that they represent the African race of this widespread species.

The specimens described here from Olduvai and Koobi Fora are almost certainly related to earlier finds from these places, which will be discussed below and in chapter 4.

Finally, mention should be made of a crushed mandible found by Broom and Robinson in 1949 at Swartkrans in South Africa (see p. 177). This was originally named *Telanthropus capensis*, but recent research has made it clear that it is most appropriately considered a South African representative of *Homo erectus*.

The evidence of *Homo erectus* in Europe is no less important. It consists of four well-known specimens that will be described in order of discovery. The famous mandible from the Mauer sands, near Heidelberg, found in 1907, was originally named *Homo heidelbergensis*, but today it seems not only unjustified to maintain a separate species for it (which implies an entirely separate European lineage, distinct from other *Homo erectus* finds discovered above), but it can reasonably be considered a member of the European race or subspecies of *Homo erectus*. On stratigraphical and faunal grounds it is believed to date from a warm

phase of the Middle Pleistocene and is usually considered to be of
Inter-Mindel or Günz-Mindel age: so although it may be a
contemporary of Pekin man it is more likely to be older. Although
the dentition of this specimen is smaller in relation to the
mandible than we see in Pekin man (and this is its main
peculiarity), the massive construction of the mandible, together
with the conformation of the symphyseal region and the vertical
ramus of the jaw, recalls very closely some of the known
specimens of *H. erectus*. The antiquity of the jaw, and its
association with animal bones broken as if by design (like those
found in deposits containing the remains of Pekin man), provide
some further argument for allocating it to the species *Homo
erectus*.

The second important discovery from Europe was made in 1960
when a group of Greek villagers found a beautifully preserved
cranium deep in the Petralona cave near Salonika, northern
Greece (Stringer 1974*a*). The stalagmitic deposit upon it and
around it could not be easily dated, and the skull was first
thought to be of late Middle Pleistocene age; but is now believed,
on the basis of the associated fauna, to be perhaps as old as the
Günz-Mindel interglacial. It is not at present possible, however,
to be precise about its date. Morphologically, it seems to carry a
mixture of characters—some characteristic of early *H. sapiens*,
some more reminiscent of *H. erectus*. The cranial vault is long
and low, as in Pekin man, but also somewhat similar to the
Neandertal and Broken Hill skulls; the endocranial capacity is
about 1,220 cc. The face is massive, and its orientation differs
from the Neandertal face; it is close to *H. erectus* (e.g., S4, S17).
The palate is extremely broad: the mandible must have been
broader than all known *H. sapiens* jaws and even the Mauer
mandible. It matches most closely the *H. erectus* S4 palate and
the Ternifine mandibles, described above. The Petralona teeth,
on the other hand, are relatively small and in this feature recall
the Mauer mandible in a very striking way.

In its morphology, this remarkable skull, which is almost
perfectly preserved, seems to be either a late *H. erectus* with a
rather advanced vault and reduced dentition (another modern

character) or an early *H. sapiens* with a very primitive face. It has
been included here in the species *H. erectus* on the basis of its
total morphological pattern, but this classification will be
strengthened if its age is confirmed to be greater than the Mindel-
Riss period. Indeed, it is more likely to be the same age as the
Mauer jaw, with which (although it has no mandible) it seems to
have much in common.

The next European find of importance was made in 1965 in a
travertine quarry in Hungary, outside Vértesszöllös (Thoma
1966, 1969). This consisted of a human occipital bone together
with four deciduous teeth. The bones and teeth came from a layer
of culture-bearing deposits, with a pebble industry of chopping
tools and with hearths. The age of the level is well established as
Inter-Mindel. It is therefore the same age as or possibly a little
more recent than the Mauer mandible. The importance of this
bone lies in the fact that in spite of its considerable thickness (a
character typical of *H. erectus*) and its wide, flat nuchal area, it is
somewhat less angular and more curved than other occipitals of
H. erectus and carries a much less pronounced occipital torus.
What is more, it is of such a size and shape as to imply with some
certainty a cranial capacity of at least 1,350 cc, which is far larger
than that of other *H. erectus* skulls and in fact the average for
modern man.

This bone, slight as it is, is a morphologically extreme specimen, considering its well-established antiquity. It seems to be far
in advance, in endocranial capacity and morphology, of the
material of similar antiquity from other parts of the world, such
as the population of Choukoutien. It is morphologically not
greatly different from the occipital of the Swanscombe skull,
which is also very thick.

The most recent addition to the European group of *H. erectus*
fossils is a frontal and other fragments from a site near Bilsrings-
leben, in East Germany, which is claimed to be of late Mindel or
early Mindel-Riss age.

These European finds are reminiscent of the skull from Salé
described above. All of them carry many characters associated

with the early *H. sapiens* fossils, especially in the occipital region and in the dentition. They are best viewed as advanced *H. erectus*, at least until their morphology and age are better known.

The Species *Homo erectus* and Its Relationship to *Homo sapiens*

The recognition of the species *H. erectus* has been based on the skeletal remains of a number of individuals that, as we have seen, show a moderate degree of individual and geographic variation. But they make up a group whose morphological characters are held by most anthropologists to be sufficiently consistent and distinctive to justify their separation as a distinct species of the genus *Homo*. If this is so, it is a matter of some importance (particularly for future reference in paleoanthropology) to provide at least a provisional diagnosis of the species, and we suggest the following definition.

> *Homo erectus*—a species of the genus *Homo* characterized by a cranial capacity with a mean value of about 1,000 cc; marked platycephaly, with little frontal convexity; massive supraorbital tori; pronounced postorbital constriction; opisthocranion coincident with the inion; vertex of skull marked by sagittal ridge; mastoid process variable, but usually small; thick cranial wall; tympanic plate thickened and tending toward a horizontal disposition; broad, generally flat nasal bones; heavily constructed mandible lacking a mental eminence; teeth large, with well-developed basal cingula; canines sometimes projecting and rarely slightly interlocking, with small diastema in upper dentition; first lower premolar bicuspid with subequal cusps; molars with well-differentiated cusps sometimes complicated by secondary wrinkling of the enamel; second upper molar may be larger than the first, and the third lower molar is variable; pelvis, at least in some races, very robust, with pronounced iliac pillar; limb bones not easily distinguishable from those of *H. sapiens* but femora with pronounced platymeria and a low point of minimum diameter.

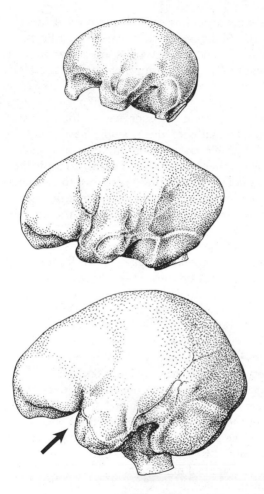

Fig. 16. Lateral view of the endocranial casts of *A*, a male gorilla; *B, Homo erectus* (from Choukoutien); and *C, Homo sapiens* (the Sylvian notch is indicated by an arrow). Approximately one-third natural size.

It will be noted from this definition that the distinction between *H. erectus* and *H. sapiens* is clear. The West European Neandertal and other Neandertaloid races do show some superficial resemblances to *H. erectus*, but when the thickness and capacity of the skull and the relatively primitive dentition of *H. erectus* are taken into account there is little likelihood of confusion. The main problem in this connection lies in the status of the Solo population. The skulls of these people have much in common with those of *Homo erectus*, and the cranial capacity is not as large as it is in other Neandertaloid races. Indeed, the age of the specimens is not clear, and they may prove to be somewhat older than has been implied. However, this population is probably not of central importance in our analysis of the course of man's evolution.

A definition of the kind given above hides a further problem. It gives the impression that populations of *Homo erectus* are all more or less centered on an average "type," which we have defined. This, however, is obviously not the case. Not only are there considerable differences between the various and far-flung geographic races, but the long period of time, perhaps 1,000,000 years, through which *Homo erectus* evolved, witnessed considerable changes in its morphology. Deep in the Djetis levels of Java lay that very primitive, heavily built skull (S4) with its 750 cc cranial capacity, while up in the Mindel deposits of Eastern Europe, dated not much more than 400,000 BP, that extraordinarily modern-looking Vérteszöllös occipital bone was discovered with its modern-sized brain. This may make trouble for the taxonomist, but it is precisely the kind of situation that paleoanthropologists are looking for: the fossil evidence of an evolving, changing lineage; one that moves from a more primitive to a more advanced state over a known period of time. The data may not respond to neat classification and categorization, but in this instance they demonstrate quite conclusively the general course and nature of hominid evolution.

The relationships of *H. erectus* are obviously a matter of the greatest interest for the problem of the origin of our own species. There is indeed a general consensus that *H. erectus* stands in an ancestral relationship to *H. sapiens*. This does not mean (as we have already emphasized) that any one population of this species was itself the actual ancestral group—it means, at the most, that the species as a whole was probably ancestral to the later species; if so, of course, the transition from one to the other may have occurred in almost any part of the Old World (though it would be possible to present a case for Europe and North Africa as the area of greatest advance). The evidence for such a hypothesis is dependent on the following line of reasoning: (1) The morphological characters of *H. erectus* conform very well with theoretical postulates for an intermediate stage in the evolution of later species of *Homo* from still more archaic types approximating the presumed common ancestral stock of the Pongidae and Hominidae. These theoretical postulates are based on comparative anatomical studies of the modern representatives of these two subdivisions of the Hominoidea and by analogy with the known evolutionary history of other mammalian groups. (2) The existence of *H. erectus* in the early part of the Pleistocene, antedating any of the well-authenticated fossil remains of *H. sapiens*, provides it with an antiquity that conforms well with the suggested ancestral relationship. (3) *H. erectus*, early *H. sapiens* populations such as that represented by the Steinheim skull, and modern types of *H. sapiens* of the Upper Pleistocene provide a temporal sequence that appears to illustrate a satisfactorily graded series of morphological changes leading from one type to the other. These arguments are precisely similar to those that have, for example, led to the conclusion that the genera *Merychippus* and *Pliohippus* were ancestral to *Equus*. But, whereas the fossil record of the Equidae has now accumulated in such detail that this phylogenetic sequence is as well demonstrated as any is ever likely to be on the evidence of paleontology, fossil hominid material is still relatively scanty. Thus the inference that the species *H. erectus* was ancestral to *H. sapiens* must be accepted for the present as a working hypothesis. But it is a working hypothesis that is fully consistent with the evidence so far available.

A further problem follows from this. If this hypothesis is

correct the species *Homo erectus* is a paleospecies that in time
gave rise to *Homo sapiens* by continuous change (see p. 25). This
being the case, it cannot be expected that it will be possible to
classify without doubt those fossils that lie near to the *H. erectus-
H. sapiens* boundary. The differences here fade to nothing, and
as more fossils from this period are discovered, more controversy
about their status will follow. It seems to be generally accepted
that the species *Homo sapiens* can be considered to date from
near the beginning of the Mindel-Riss interglacial, ca. 300,000
BP. It seems best therefore to consider human fossils that are
older than this time boundary to belong to *H. erectus* and those
younger to be members of *H. sapiens*.

Having said this, it should be clear that definitions such as have
been given can be no more than a general guide to a typical
member of the paleospecies concerned, and that fossils will
undoubtedly be found that lie so close to the temporal borders of
the species that their status will always remain *morphologically*
uncertain. The fossil skulls from Europe (Vértesszöllös and
Petralona), and from Salé in Morocco, discussed here as members of the species *Homo erectus*, fall into this "uncertain"
category.

During the earliest times of *Homo erectus*, the most primitive
forms doubtless did evolve in turn from their predecessors, and
so, in the following discussion, this same problem will arise again.
That the problem exists at all is due only to the increasingly well
documented fossil record that now exists of the evolving Hominidae.

If the thesis is correct that the Hominidae and the Pongidae are
divergent radiations from a common ancestral stock—the result,
that is to say, of a phylogenetic dichotomy—the evolutionary
precursors of *H. erectus* must presumably have shown morphological characters approximating much more nearly an anthropoid level of evolution. To some extent these characters might be
tentatively predicated by a consideration of comparative anatomical data, by extrapolation backward of the *H. sapiens-H.
erectus* sequence, and by analogy with the sort of morphological
gradations that are known (from more complete fossil records) to

have occurred in the evolution of other mammalian groups. For example, it is evident enough that one of the main features of hominid evolution (at least in its later stages) has been the progressive development of the brain; and there is some evidence that in the Early and Middle Pleistocene it proceeded with unusual rapidity in comparison with evolution rates in general (see Haldane 1949). It may be presumed, therefore, that in the immediate ancestor of *Homo erectus* the cranial capacity would certainly have been still smaller—perhaps, indeed, not very much greater than that of the largest anthropoid ape of today, the gorilla. With this would probably have been associated massive jaws and large teeth similar to those of *H. erectus*, but of somewhat more impressive proportions, and, in further correlation, powerful temporal and masseter muscles with extensive areas of bony attachment to the skull. Since the essential characteristic features of the hominid dentition (by which it is so strongly contrasted with the pongid type of dentition) were already well established in *H. erectus*, it might be expected that in the immediately ancestral species these hominid features would also be clearly evident (in spite of the size of the teeth), though no doubt more primitive in certain details. Last, the fact that in *H. erectus* the limb bones had already fully achieved the morphology and proportions of *H. sapiens* strongly suggests that the evolutionary modifications of the limbs related to erect bipedalism had been acquired very early in the line of hominid development and would thus already be apparent in the ancestral species. These inferences are, of course, based on indirect evidence and can be verified only by the discovery of actual fossil remains. The remarkable fact is that they actually have been verified to a large extent by the discovery of the earliest *Homo* and australopithecine fossils in Africa.

Homo habilis

During the past fifteen years, a small number of hominid fossils have been discovered that appear to stand morphologically between *Homo erectus*, as we have defined it, and the ancestral genus *Australopithecus*. The first of these to be recognized as such was a group of fragmentary bones including two parietals, a mandible, and some phalanges found by Louis and Mary Leakey at Olduvai Gorge in Tanzania (see p. 132). These specimens carried some unusual characters not found in *Homo erectus* or *Australopithecus*, and for this reason they were named *Homo habilis* by Leakey, Tobias, and Napier (1964). Together with this type specimen (OH7) were grouped a number of other specimens from Olduvai (listed in table 2). The creation of a new species on the basis of such limited evidence was originally considered unjustified by Campbell and Le Gros Clark, but further recent discoveries of teeth and a fragmentary cranium at Omo (Boaz and Howell 1977) and of skulls and other fragments from the Lower Koobi Fora Formation at Koobi Fora (R. Leakey 1973; Leakey and Leakey 1978) support recognition of this taxon. Finally, it has now been well established that an early population of *Homo* is also represented at the South African sites Sterkfontein (member V) (Hughes and Tobias 1977) and Swartkrans (member I) (see p. 177).

These earliest forms of the genus *Homo* can be distinguished from *Homo erectus* by a number of characters: the skull is very

Table 2 Important Specimens
from Africa Attributed to the
Species *Homo habilis*

Site	Specimen	Age (MYBP)
Olduvai	OH 4, 6, 7, 8, 13, 14, (16), 24, (35)	1.8–1.6
Omo	L. 894.1	1.84
Koobi Fora	KNM-ER 1470, 1481, 1501, 1503, 1590, 1802	2.1–1.8
Sterkfontein (V)	Stw 53	2.0–1.5
Swartkrans (I)	SK 847, 45, (27), 85	2.0–1.5
Associated remains from Java		
Modjokerto	M1	ca. 1.9
Sangiran	S1, 4, 5, 6, 9, 22	ca. 1.9

Note: The association of specimens in parentheses is uncertain.

thin, with moderate brow and smoother, more rounded form, and in this character is reminiscent of the Modjokerto child (p. 99); the cranial capacity is substantially lower than that of *Homo erectus* and ranges from ca. 600 to 750 cc (with a sample of five skulls); the incisors are broad, the canines small; the premolars are recognizable by their relatively long mesiodistal diameter; both they and the molars are larger and more molarized than those of *Homo erectus*. Some postcranial fragments are known from Olduvai Gorge (bed I) including a tibia and fibula and part of a foot skeleton. The tibia and fibula (OH 35) (which are possibly but not certainly part of the *H. habilis* population) have been described as modern in form. The foot skeleton (OH 8) was found in close association with the type mandible (OH 7). The foot is almost complete except for the phalanges, the distal ends of the metatarsal bones, and the posterior part of the calcaneus (fig. 17). This is the most complete fossil hominid foot we have until the time of Neandertal man: it is modern in the most important respects; namely, that the metatarsal of the great toe is stoutly built, as in modern man, and is not divergent (as shown by the presence at its base of an articular facet for the second metatarsal) (Day 1976). The fifth metatarsal is also the second most robust of the foot, as it is in modern man. There is also evidence of a transverse as well as a longitudinal arch in the foot, as found in modern man. A femur and a group of foot bones from the Lower Koobi Fora Formation are also described as very similar to their modern counterparts (R. Leakey 1973; Lovejoy 1978).

An important group of thirteen hand bones was also present at Olduvai, most of which are juvenile (as is the mandible OH 7). A recent study by Day concludes that this hand had powerful fingers, flat nails, and (on the basis of a broad saddle surface on the trapezium for the first metacarpal) a thumb capable of a wide range of movement including rotation out of the plane of the palm to give a precision grip (Day 1976).

These bones suggest a population of small, lightly built people and are rather different from the heavily muscled bones of *Homo erectus*. Two crania and some other remains from Koobi Fora,

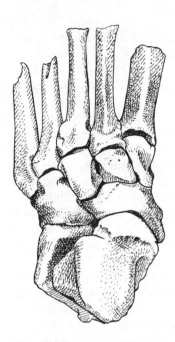

Fig. 17. Dorsal view of the foot skeleton found at Olduvai. The front ends of the metatarsals and the back part of the calcaneus are missing. Approximately three-quarters natural size. From a photograph by Dr. M. H. Day.

however, suggest a somewhat heavier form, intermediate between the Olduvai specimens and *Homo erectus*.

As already indicated, the taxonomy of this group presents problems owing to the sparseness of fossils. An assessment of the present African evidence does indicate the existence there of populations quite distinct from those of *Homo erectus* that we have discussed, and closer to the ancestral genus *Australopithecus*. Some authors have classified these fossils as *Australopithecus habilis*, thus recognizing that the *Australopithecus–Homo* transition took place about 1.5 million years ago. It is a convention of zoology, however, that genera distinguish groups of species with a fairly uniform adaptive facies (Simpson 1961), that is to say, species that share a broadly similar pattern of adaptation. Since the most striking human adaptation is man's culture, it is therefore significant that unlike *Australopithecus*, *Homo habilis* is associated with stone tools and does indeed appear to be the earliest maker of a stone industry to a regular pattern. The earliest industry is recorded at Omo (ca. 2.0 million years BP), but here it is not associated with hominid bones. Stone industries dating from about 1.8 million years BP occur with early *Homo* at Sterkfontein, Swartkrans, and Olduvai.

The difficulty of assigning these early fossils to a recognized taxon is further increased by uncertainty about their relationship to the *Homo erectus* material from the Djetis levels in Java, discussed above. This material has always been broadly classified as early *Homo erectus*, but Tobias and von Koenigswald (1965) have shown that it carries many features that are paralleled by the *Homo habilis* fossils from Olduvai. If these Javan forms prove in due course to be of the same age and general morphology as the African specimens—and turn out to represent the Asian members of this species—then the name given to the Modjokerto child by von Koenigswald, *Homo modjokertensis*, will have priority over *Homo habilis* because it was published in 1936.

For the present, we have retained the well-known term *Homo habilis* for these intermediate specimens. As the science of paleoanthropology develops, revision of this nomenclature may well be required.

Any diagnosis of this species is bound to be very provisional, owing to the limited fossil evidence from this period:

> *Homo habilis*—a species of the genus *Homo* distinguished mainly by its small cranial capacity of mean value ca. 670 cc; a relatively thin skull, smooth and with moderate brow; lightly built face, with considerable alveolar height; mandible light, without chin; the incisors broad, canines small; premolars with long mesiodistal diameter; both molars and premolars large, long, and somewhat molarized. Postcranial limb bones suggest small erect manlike creatures no more than 4 to 4½ feet in

height. The leg and foot do not appear to fall significantly outside the morphological range of variation found in modern man.

The Genus *Homo*

We have attempted to give a diagnosis of the species *H. sapiens*, *H. erectus*, and *H. habilis* on the basis of skeletal characters, so as to provide some standard of comparison for further paleontological studies. For the same reason it is also desirable to provide a formal diagnosis of the genus *Homo*:

> *Homo*—a genus of the family Hominidae, distinguished mainly by a large cranial capacity with a mean value of more than 1,100 cc but with a range of variation from about 600 cc to over 2,000 cc; supraorbital ridges variably developed, moderate in *H. habilis*, becoming massive in *H. erectus*, showing considerable reduction in *H. sapiens*; facial skeleton orthognathous or moderately prognathous; occipital condyles situated approximately at the middle of the cranial length; temporal ridges variable in their height on the cranial wall, but never reaching the midline to form a sagittal crest; mental eminence sometimes well marked in *H. sapiens* but absent in *H. erectus* and *H. habilis*; dental arcade evenly rounded, usually with no diastema; first lower premolar bicuspid; molar teeth rather variable in size, with a relative reduction of the last molar in later forms; canines relatively small, with no overlapping after the initial stages of wear; limb skeleton adapted for a fully erect posture and gait.

The geologic antiquity of the genus *Homo*, on the basis of the fossil record discussed in this chapter, goes back in Asia to ca. 1.9 and in Africa to at least 1.8 million years BP. Evidence at Koobi Fora suggests an earlier date, and Boaz and Howell (1977) select a figure of 2.3 million years BP as the age of the earliest members of the genus *Homo* on the basis of the present evidence.

Four

Australopithecus

The Provenance and
Age of *Australopithecus*:
South Africa

In 1925 Professor R. A. Dart, of the Witwatersrand University in Johannesburg, described the well-known Taung skull from Cape Province as representing a new type of hominoid, to which he gave the generic name *Australopithecus* (Dart 1925).[1] He pointed out that, in spite of the obviously apelike proportions of the braincase, the relatively large jaws, and the pronounced prognathism, there are certain features (particularly in the dentition) in which this fossil hominoid approximates more closely to the Hominidae than any of the known anthropoid apes do. In general, other anatomists regarded the morphological resemblances as little more than interesting examples of parallelism having no particular reference to hominid evolution. In 1936 and the following years, the late Dr. Robert Broom (at that time working at the Transvaal Museum, Pretoria) discovered at three sites near Johannesburg (Sterkfontein, Kromdraai, and Swartkrans) much more numerous and complete remains of similar hominoids, comprising many skulls, several jaws, parts of the limb skeleton (including two well-preserved specimens of the innominate bone), and several hundred teeth. In these excavations and in the study of the fossil material he was assisted by Dr.

J. T. Robinson. Broom died in 1951 at the age of eighty-four; for several years Dr. Robinson was responsible for continuing the excavations and for making still further important discoveries. At another site altogether, in cave deposits at Makapansgat (about 150 miles north of Johannesburg), Dart found, in 1947 and later, more remains of *Australopithecus*, including a crushed skull, an immature mandible, an occiput, parts of two pelvises, and a maxillary fragment. Work was resumed at Swartkrans in 1965 under the direction of Dr. Brain, Broom's successor at the Transvaal Museum. Further excavation was also undertaken at Sterkfontein by Tobias and Hughes from late 1966. Both sites have again yielded important hominid remains, and together all these sites demonstrate the extent in time and space of these early hominids in the Transvaal.

With his collaborators, Broom described his remarkable collection of fossils in a number of separate papers and in three monographs (Broom and Robinson 1952; Broom, Robinson, and Schepers 1950; Broom and Schepers 1946). The material now available is so abundant that it will certainly take many years before anything like an exhaustive account of it can be completed. Broom himself recognized this well, and in his monographs published by the Transvaal Museum he attempted little more than to provide a general survey of his discoveries, illustrated by photographs and drawings. But his descriptions made it very evident that Dart's first appraisal of the Taung skull was basically correct—that is, that morphologically *Australopithecus* approximates the Hominidae much more closely than any of the known anthropoid apes do (whether Recent or extinct) and that the genus as a whole may well include the ancestral stock from which *Homo* was derived (or that at least it was very closely related to the ancestral stock). This conclusion was later reinforced by a detailed and careful study of the australopithecine dentition by Robinson (1956). The history of these discoveries in South Africa has already been recounted more than once, and there is thus no need (except incidentally) to repeat it again here (see Clark 1967).

All the australopithecine remains from the Transvaal have

been derived from cave deposits or fissures in formations of dolomitic limestone and in most cases were firmly imbedded in a dense stalagmitic breccia matrix. Stratification was very difficult to recognize, and Broom believed that the cave deposits were laid down rather rapidly over a short period. It has now proved possible to distinguish a series of superposed stratigraphic horizons at all the sites except Taung, which has been destroyed by mining. It has, however, proved impossible to date the material radiometrically. As a result, paleontologists have attempted to get some idea of the absolute age of the sites by comparing their Pliocene and Pleistocene fauna to faunas found 2,000 miles to the north in East Africa; faunas that carry well-established potassium-argon dates. These studies have progressed slowly; with such a distance separating the dated from the undated faunal assemblages, there is no reason to suppose that they should be exactly equivalent at any time. As Cooke (1952) has put it, "it seems that this southern tip of the African continent was in the nature of a cul-de-sac in which archaic forms survived long after they had become extinct in the more rigorous and variable climate of Quaternary Europe." However, some consensus is at last developing as a result of very intense studies, that have established on the basis of the accompanying fauna, especially the bovids (Vrba 1975), the relative ages of the fossils at each site (see fig. 18). It seems now to be generally agreed that the South African sites span from about 3.0 to less than 1.0 million years BP, as indicated in figure 18. Apart from the indications they may provide of geologic age, some of the mammalian remains in the australopithecine deposits provide very important ancillary evidence of the climatic environment at the time, for they confirm the geologic evidence that it was not very different from the present time, that is to say, a climate tending to aridity—it certainly was not a region of tropical forest, such as is suitable for arboreal anthropoid apes as we know them today. There is faunal evidence that the bush cover may have been greater during the earlier (Sterkfontein) time span than the later (Swartkrans) span, when open grassland predominated, as it does today.

East Africa

In 1913, a human skeleton associated with a Pleistocene fauna was excavated at Olduvai Gorge in Tanzania by the German anthropologist Hans Reck. It was not until 1931 that Louis Leakey returned to the site and began the systematic study of the fauna and extensive stone industry. Soon after this the Olduvai skeleton was shown to be an intrusive burial from a much higher level than that containing the stone tools and Pleistocene fauna. During the next twenty-eight years, in spite of continuous searching, Leakey found only two parietal bone fragments and two teeth that appeared to be hominid. It was in 1959 that Mary Leakey, his wife, made the first great discovery from Olduvai—a skull of a very robust *Australopithecus*—on an ancient land surface or "living floor," together with numerous bones that proved to be food remains and stone tools and waste flakes in abundance (Tobias 1967, M. Leakey 1971). Further discoveries were made in the following years, and today we have remains of more than fifty hominids from the gorge, part of which have been classified as *Australopithecus* (Oakley, Campbell, and Molleson 1971-77).

The sedimentary deposits through which the gorge has been cut are interbedded with aerial tuffs from nearby volcanoes that have proved most suitable for dating by the potassium-argon method. As a result we now have an idea of the absolute age of these deposits together with their fauna and industries: they span the period from 1.8 million years BP to 0.2 million years BP. The remains that concern us in this chapter were found in bed I and lower bed II—dating from between 1.8 and about 1.5 million years BP (see fig. 18).

In 1966 an international expedition under the direction of Professors F. C. Howell and Y. Coppens initiated work along the west bank of the Omo River in southern Ethiopia. During the following nine years they collected and excavated nearly two hundred teeth and nineteen bone fragments including three partial crania that are from deposits well dated by potassium-argon between 3.0 and nearly 1.0 million years BP (Howell and Coppens 1974). The majority of these have been classified as *Australopithecus*.

In 1968 Richard Leakey led an expedition to explore the desert regions near Koobi Fora to the east of Lake Turkana, in northern Kenya. In the past eight years more than one hundred well-preserved hominid fossils have been recovered, mostly as a result of surface collection (Leakey and Leakey 1978). Among them are

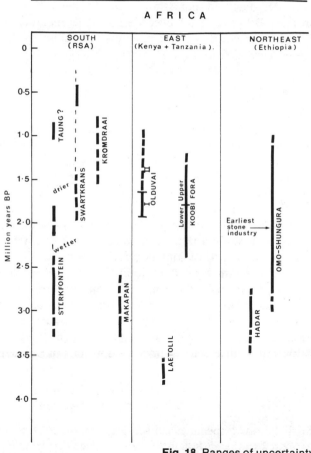

Fig. 18. Ranges of uncertainty in the ages of important *Australopithecus* fossils.

some extremely well-preserved crania of *Australopithecus* as well as of *Homo*, derived from the upper and lower members of the Koobi Fora Formation. Potassium-argon determinations indicate an age of between 2.0 and 1.5 million years BP.

At Laetolil, twenty-five miles south of Olduvai Gorge, Dr. Mary Leakey has, since 1974, discovered more than thirteen hominid fragments—a maxilla, mandibles, and teeth. They all lie between tuffs dated by potassium-argon to be 3.77 and 3.59 million years BP. They have been described as members of the genus *Homo* but clearly carry characters very close to those of *Australopithecus* (White 1977).

Northeast Africa

In 1973 an international expedition began exploration of the Afar triangle in northeast Ethiopia. At a site called Hadar, a number of very important hominid fossils have since been collected, which lie above and below a tuff dated by potassium-argon at 3.0 million years BP (Johanson and Taieb 1976). Some of these specimens have been attributed to *Australopithecus*, and they will all be discussed later in this chapter.

This brief survey of some very extensive discoveries throughout eastern Africa is summarized in figure 18. It is to be noted that although the sites in East and northeast Africa all carry well-established potassium-argon dates (only the Koobi Fora dates are somewhat controversial), the South African sites are still dated only by faunal comparison over a long distance. It seems probable, however, that the dates given in the figure are good approximations. The anatomy and classification of these numerous fossils will be discussed in later sections of this chapter.

The Nomenclature of the Australopithecinae

The fossil hominid specimens discovered at these sites are all closely related to the early South African discovery from Taung, though in their publication they have been classified into a number of different genera and species as follows:

Taung	*Australopithecus africanus* Dart, 1925
Sterkfontein	*Plesianthropus transvaalensis* Broom, 1936
Kromdraai	*Paranthropus robustus* Broom, 1938
Swartkrans	*Paranthropus crassidens* Broom, 1949
Makapan	*Australopithecus prometheus* Dart, 1948
Olduvai	*Zinjanthropus boisei* Leakey, 1959
Omo-Shungura	*Paraustralopithecus aethiopicus* Arambourg and Coppens, 1968

Fossil hominid specimens from the other sites not listed above were placed in one or more of the above species or simply described as either *Homo* or *Australopithecus*. This multiplication of taxonomic terms undoubtedly tended to sidetrack the main issues in the controversies that followed the first announcements of the discoveries, and it was probably also responsible for some of the skepticism certain anatomists first expressed in regard to them. Broom thought he was able to detect morphological differences that justified the recognition of four different genera, and he believed this was further supported by paleontological evidence of a considerable gap in geologic time separating one genus from another. However, as we have noted, the consensus among geologists who have examined the sites is that the time sequences are not so different as to justify generic distinctions that have only a doubtful morphological basis. But the main point of criticism is that at the time of their discovery none of the several genera and species was adequately defined in formal diagnoses, nor has it yet been really satisfactorily demonstrated that the morphological differences are in fact greater than may be explained by individual variation or sexual dimorphism or by variations in local races or geographical varieties of the same species. Such differences as do exist may justify subspecific distinctions or possibly even specific distinctions. But it appears certain that, at the most, they are no greater than those that (for example) distinguish the two closely related species of chimpanzee, *Pan troglodytes* and *P. paniscus*. Further, if, as it is now generally agreed, the morphological contrasts between *Homo erectus* and *Homo sapiens* are not obtrusive enough to permit more than a specific distinction from the latter, then by comparison it would be difficult to justify more than a specific distinction for the australopithecines from different localities. The crucial argument for a taxonomic differentiation among the South and East African fossil australopithecines must ultimately depend on (1) the demonstration of the range of variation each local group shows and (2) how far the variations compare with those that are accepted as adequate for specific or generic distinctions in other related groups of hominoids (i.e., in the Hominidae and Pongidae). The view taken here is that all the remains so far discovered from these sites represent only two genera: *Australopithecus* and *Homo*, but that they may represent a number of species. To avoid confusion, they will collectively be referred to as the Australopithecinae (or more colloquially, the australopithecines), even though it may eventually be decided that to place *Australopithecus* in a separate subfamily is carrying its taxonomic distinction too far. So far as the different local groups of the Australopithecinae are concerned, it seems better, for the present, to avoid the different generic terms that have been applied by their discoverers (apart from *Australopithecus*) and to use the place names of the local sites for reference purposes.

The Morphological Characters of the Australopithecinae

The cranial, skeletal, and dental morphology of the Australopithecinae is now known from fossil material notable for its great quantity and for the completeness and excellent state of preservation of many of the specimens. Indeed, it is probably true to say that we now know a good deal more about the anatomy of this group and its range of morphological variation than of any other group of fossil hominoids.

The Skull and Endocranial Cast

Skulls and mandibles (or portions of them) representing juveniles, adolescents, and young and old adults have been found at all the sites we have listed. The first almost perfect specimen of a skull was exposed at Sterkfontein in 1947 (Sterkfontein skull 5)— a practically complete and undistorted skull, from which, however, the mandible was missing (fig. 19). It was found embedded in a dense stalagmitic matrix and was developed therefrom with consummate skill by Broom (Broom, Robinson, and Schepers 1950). Almost complete, but severely crushed skulls have also been found at Swartkrans. The Taung specimen (which is the type specimen of the genus) consists of the facial part of a juvenile skull, almost complete and undistorted, associated with a natural endocranial cast. Still other skull relics are represented by portions of the cranium and facial skeleton from most of the sites in the Transvaal, numerous specimens of the maxilla and palate, and several mandibles. Some of this other material is rather fragmentary but nevertheless provides information of critical importance. The australopithecine skull ("*Zinjanthropus*") from Olduvai is that of an adolescent individual and is virtually complete except for the mandible (see fig. 20). Two almost perfect skulls are known from Koobi Fora together with much other cranial material.

The most obvious feature of the australopithecine skull as a whole is the combination of a small braincase with large jaws, and it is this that gives to it such a pongid appearance. It is well to emphasize these proportions from the outset, because (as well recognized by Dart, Broom, and others) they are primitive characters in which the Australopithecinae contrast strongly with *H. sapiens* or even *H. erectus*. But it is also important to note that, although the general proportions are certainly pongid in the sense that they approximate the level of development still preserved by the modern anthropoid apes, they are not necessarily of taxonomic significance for the problem of deciding whether the Australopithecinae should be allocated in a natural classification to the Pongidae or the Hominidae. For we have already noted (p. 123) that the combination of a small braincase of approximately pongid dimensions with large jaws must certainly have been a characteristic of the early phases in the sequence of hominid evolution, that is to say, after the Hominidae had become definitely segregated in their evolutionary history from the Pongidae.

What is quite clear is that both features (jaws and braincase) vary independently in size within the Australopithecinae. We have specimens with small brains and relatively small jaws (at Sterkfontein), with small brains and big jaws (at Olduvai and Koobi Fora), and with big brains and quite small jaws (Koobi Fora) classified here as *Homo habilis*, but none with big brains and big jaws. The skulls can be graded fairly easily on the basis of the size and robusticity of the masticatory apparatus (the jaws and the bony structures of the skull associated with their movement). In this sense, some skulls are extremely robust, some relatively gracile in form. The most robust are from East Africa (Olduvai and Koobi Fora), the most gracile are from northeast and South Africa (Hadar and Sterkfontein), as well as the East African sites. Some sites in South Africa carry skulls of intermediate robusticity.

The range of variation in the cranial capacity of the Australopithecinae was evidently considerable. The following figures are based on the careful work of Dr. Ralph Holloway (1975). Four specimens from Sterkfontein range in capacity from 428 to 485 cc. One from Makapansgat was ca. 435 cc. The endocranial cast of the immature Taung skull (the right half of which is almost complete) has an estimated volume of 405 cc. In this individual the deciduous dentition was still in place, and the first permanent molar had recently erupted. On the assumption that the rate of brain growth in the Australopithecinae was equivalent to that of modern anthropoid apes, it may be estimated that the cranial capacity of the adult Taung individual would have approximated 440 cc. The Swartkrans skulls and jaws are considerably larger than those found at Sterkfontein, but, unfortunately, none of the calvariae are sufficiently well preserved to permit accurate estimates of cranial capacity. A finely preserved endocranial cast, however, gives a figure of 530 cc.

From Olduvai we have only one reliable cranial capacity—530 cc for the very robust skull Olduvai 5. At Koobi Fora robust skulls remain closer to 500 cc (careful estimates of these specimens have yet to be published).[2]

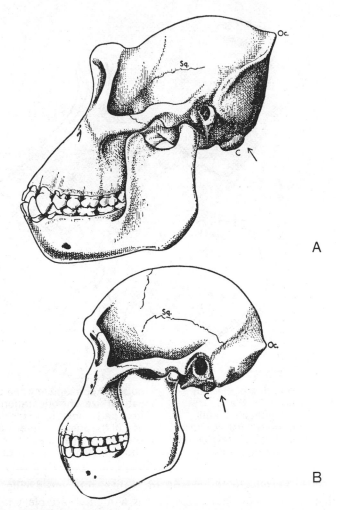

Fig. 19. The skull of *A*, a female gorilla, compared with *B*, the Sterkfontein australopithecine skull no. 5, illustrating the more lightly built type of *Australopithecus* (*A. africanus*). In the latter the mandible and dentition have been reconstructed by reference to numerous other specimens. Note the relative position of the occipital protuberance (*Oc.*), the occipital condyle (*C*), and the squamous suture (*Sq.*). The arrow indicates the axis of the foramen magnum. Approximately one-third natural size.

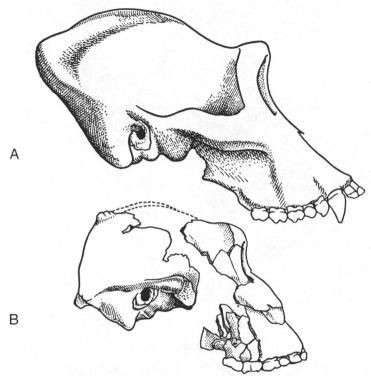

A

B

Fig. 20. Side view of the skull of a male gorilla (A) and the australopithecine skull ("*Zinjanthropus*") from Olduvai (B), illustrating the heavily built type of *Australopithecus* (A. *boisei*). Approximately one-third natural size. The orientation of the front part in relation to the back part of the australopithecine skull is slightly conjectural. From a photograph by Dr. L. S. B. Leakey.

It is extremely important to realize, in considering these figures, not only that brain size is a highly variable character within a population, but that it varies with body size, so that an increase in cranial capacity is not alone a significant taxonomic character unless it is related to body size. Thus the chimpanzee (*Pan troglodytes*) has a mean cranial capacity of 383 cc while the much smaller pygmy chimpanzee (*Pan paniscus*) has a mean of only 342 cc. There is reason to believe that some populations of

Australopithecus were no bigger than pygmy chimpanzees. In this case their cranial capacity is already considerably bigger than that of an ape of equivalent size. This is true of all the australopithecines. The very high *Gorilla* values (up to 752 cc) came from enormous animals weighing more than 400 lbs.

In some of the discussions resulting from the first discoveries of australopithecine skulls, rather unnecessary emphasis was placed on the exact value of the cranial capacity, apparently on the assumption that upon this character mainly depends the question whether the Australopithecinae are to be regarded as representatives of the Pongidae or whether they are hominids of very primitive type. This assumption, again, seems to have been based upon the supposition that there is an arbitrary point in the expansion of the brain (some have suggested a cranial capacity of 750 or 800 cc) that forms a sort of dividing line between man and ape. But (let us once more emphasize) the expansion of the brain to the size characteristic of the *later* Hominidae evidently did not occur until long after this family had become segregated in the course of evolution from the Pongidae, and a large cranial capacity is thus not a diagnostic feature of the family Hominidae as a whole. In any case, of course, we know nothing of the intrinsic structure of the australopithecine brain, which, in spite of its relatively small size, may have had a more elaborate neural organization than that of the modern anthropoid apes.

Several natural endocranial casts of the Australopithecinae have so far been found, two almost complete specimens at Taung and Sterkfontein and three fragmentary specimens from Sterkfontein and Swartkrans. Plaster casts have also been prepared of the endocranial cavity of skulls or portions of skulls found at Sterkfontein, Kromdraai, and Makapansgat. Most of these endocranial casts were studied in some detail by Schepers (Broom, Robinson, and Schepers 1950; Broom and Schepers 1946); but, in his attempt to extract every possible item of information from them (perhaps excusable, considering the unique character at that time of these South African fossils), he undoubtedly allowed himself to overstep the limits of legitimate deduction and to read into the surface details of the casts morphological and physiological conclusions that are certainly not warranted by the facts. As might be expected from a consideration of the skulls, the general size and the main proportions of the australopithecine endocranial cast do not differ markedly on superficial inspection from those of large chimpanzees or gorillas; and their surface irregularities unfortunately do not provide much information relating to those morphological details that are commonly regarded as of particular interest in contrasting the hominid with the ape cerebrum. But there are certain points of relevance that may be mentioned here. In the first place, the "Sylvian notch" between the frontal lobe and the temporal pole is deeper and more sharply angulated than it is in the generality of modern large anthropoid apes of equivalent size, as is evident enough when direct comparisons are made between the actual endocranial casts (though it has been disputed by those who have relied only on diagrams and drawings). However, this is simply a reflection of the sharp, undercut character of the orbitosphenoid in the base of the skull, which again is evidently the result of the degree of flexure of the basis cranii shown by the australopithecine skull.

The second feature is the richness and complexity of the convolutional impressions on the endocranial casts. Although this character has been overemphasized by some writers, it is actually quite obtrusive. But comparisons with endocranial casts of modern anthropoid apes, *H. erectus*, and *H. sapiens* are not easy to make, for in all these types the convolutions do not imprint themselves on the endocranium nearly so distinctly. A direct visual comparison of the fossil casts with the actual brains of the large apes suggests that in the Australopithecinae the convolutional pattern of the cerebral hemispheres was probably more complex. Some emphasis has been laid on the posterior position and reduced size of the "simian" or lunate sulcus in the occipital lobe of the brain, features that may in some sense be regarded as hominid, since they indicate an expansion of the parietal cortex to a degree not found in the modern apes.

Beyond this, the inferior surface of the parietal lobe and the inferior frontal gyrus also appear to be expanded above the size seen in apes, and the shape of the temporal lobe and the orbital surface of the frontal lobe are more similar to those of later hominids than to living pongids.

It will be apparent from what has been said that, taken by themselves, the endocranial casts of the Australopithecinae do not permit firm statements regarding the convolutional pattern of the brain itself. But such indications as there are do assume significance when taken in conjunction with all the other anatomical characters of the skull, dentition, and limb skeleton.[3]

It has already been mentioned that the australopithecine skull is remarkably primitive in its *general* proportions, related to the combination of a small braincase with relatively large jaws. This character is associated with a remarkable assemblage of morphological details in which, in combination, the skull contrasts strongly with that of the Pongidae and is to be matched only in the Hominidae. These hominid features may be listed as follows.

a) The Cranial Height. One of the obtrusive features of the exceptionally well preserved australopithecine skull (Sts 5) from Sterkfontein is the height of the cranial vault above the level of the orbit (figs. 19 and 21). This has been expressed by the "supraorbital height index," which measures the height of the braincase above the level of the supraorbital margin in relation to the total height above the Frankfurt plane (Le Gros Clark 1950). A comparison with a series of gorilla, chimpanzee, and orang skulls shows that in Sterkfontein 5 the relative height exceeds the range of variation in anthropoid apes and actually comes within the range of hominid skulls.[4] The figures for this index are listed in table 3 (p. 146). The high cranial vault in the australopithecine skull, it should be noted, is related not to the actual size of the braincase but to the fact that it is set higher in relation to the upper part of the facial skeleton. This again is evidently related to the flexure of the basicranial axis, which, being more marked than it is in the anthropoid apes, raises the braincase to a relatively higher level.

b) The Low Level of the Occipital Torus and Inion. This has been estimated on the basis of the nuchal area height index in Sterkfontein 5 (figs. 19 and 21, table 3). It shows that the

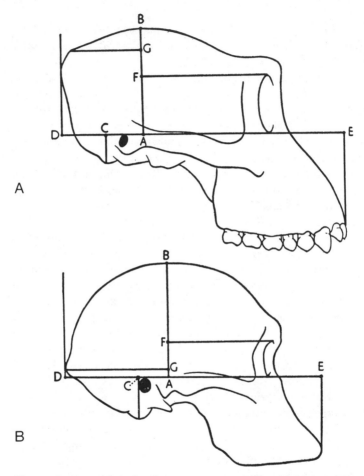

Fig. 21. Outlines of skulls of *A*, a female gorilla, and *B*, *Australopithecus* (Sterkfontein 5). The indexes mentioned in the text are illustrated here: nuchal-area height index *AG/AB*; supraorbital height index, *FB/AB*; condylar-position index, *CD/CE*. Approximately one-third natural size.

occipital torus (in relation to the Frankfurt plane) occupies as low a level as it does in skulls of *H. sapiens*, and the index is far below the range of variation shown in the large anthropoid apes. As is well known, in adult apes the torus forms a crest reaching high up on the occipital aspect of the skull, thus extending very considerably the nuchal area for the attachment of the powerful neck muscles. On the other hand, in the Australopithecinae (as in *Homo*) the nuchal area is, by comparison, very restricted and, moreover, faces downward rather than backward on the occipital squama. That this was a consistent feature of the australopithecine skull is indicated by the appearance of the occipital region in other specimens. For example, the Olduvai 5 skull and the well-preserved occipital bone of *Australopithecus* found at Makapansgat also show quite a small nuchal area, limited above by a comparatively weak torus. An examination of the crushed Sterkfontein 1 and the fragmentary Kromdraai skull provides similar evidence.[5]

c) The Position of the Occipital Condyles. One of the features in which the hominid skull contrasts strongly with the pongid skull is the forward position of the occipital condyles in relation (1) to the total skull length and (2) to the transverse level of the auditory apertures. In the Pongidae, in association with the different poise of the head during life, the condyles are situated a considerable way behind the midpoint of the cranial length and also well behind the auditory apertures, and it is interesting that the latter relationship holds good for the immature as well as the adult skull. In the Australopithecinae the relative position of the condyles has been accurately determined in Sterkfontein 5, and on the basis of the condylar-position index it can be shown that they are definitely farther forward in relation to the total cranial length than in adult orangs and chimpanzees[6] and the great majority of adult gorillas (fig. 21; table 3). In the latter a few specimens have been found in which the condylar-position index equals or even exceeds that of the Sterkfontein skull. But in these instances the indices are not strictly comparable, for in large male gorillas the sagittal crest commonly extends backward beyond the braincase as an attenuated flange, which thus exaggerates the total skull length when the latter is expressed by the usual overall measurement (Clark 1952). In any case, it is important to recognize that the position of the condyles in relation to the skull length does not provide more than an approximate measure of the balance of the head in relation to the carriage of the body—to obtain such a measure accurately, we should require (inter alia)

Table 3 Cranial Indices

Species	Supraorbital Height (100 FB/AB)	Nuchal Area Height (100 AG/AB)	Condylar Position (100 CD/CE)
Gorilla gorilla	48	69	24
Pan troglodytes	47	52	23
Pan paniscus	—	—	35
Dryopithecus africanus	55	33	30
Australopithecus africanus	61	8	39
A. boisei	52	9.5	55
Homo erectus	63–67	33–56	—
H. sapiens	64–77	1–16	77–81

Note: For discussion, see text, pp. 143–46.
Source: Data from Rosen and McKern (1971), Tobias (1967).

to know the position of the center of gravity of the whole head during life. It is also to be noted that, although the high condylar-position index indicates a relatively forward position of the condyles in the Sterkfontein skull, in relation to the total skull length they are not situated as far forward as in *Homo* because of the massive and prognathous jaws. On the other hand, their forward position relative to the auditory apertures (and also to other elements of the skull base, such as the extremity of the petrous bone and the carotid foramen) parallels that of *Homo*. This important relationship is consistently present in all the known australopithecine skulls in which the cranial base is sufficiently well preserved. Finally, apart from their relative position, it is to be observed that in the australopithecine skulls the long axis of the occipital condyles is approximately horizontal (in reference to the Frankfurt plane), as in *Homo*, and does not slope upward and backward as it normally does in pongid skulls of equivalent size and massiveness.[7]

The three features of the australopithecine skull to which attention has just been drawn are evidently all related directly or indirectly to a common factor—the poise of the head in relation to the vertebral column. The small nuchal area and the low position of the occipital torus provide clear evidence that the neck musculature was not extensively developed to hold up the head in relation to a forward-sloping cervical spine, as it is in the modern large anthropoid apes. The forward position of the occipital condyles is associated with an increased flexure of the basicranial axis, and this flexure has also led to an upward displacement of the braincase relative to the facial skeleton, with a resultant increase in the cranial height. If this interpretation is correct, it is particularly interesting to note that a statistical analysis has shown that the three indices—supraorbital height, nuchal area height, and condylar position—taken in combination, definitely place the australopithecine skull outside the limits of variation of the large anthropoid apes and indicate a remarkable approximation to *Homo*. And if due consideration is given to their significance, it is a reasonable, and indeed an obvious, inference that the bodily posture of the Australopithecinae approximated that characteristic of *Homo* and was very different from that of the Pongidae. As we shall see, the validity of this inference is confirmed by the anatomy of the pelvis and limb bones.

d) Mastoid Process. In all the available australopithecine skulls in which the mastoid region is sufficiently well preserved, there is a well-marked pyramidal process of typical hominid form. Indeed, it is already well developed in the two immature Swartkrans skulls (of individuals equivalent in dental age to seven and eleven years in modern man). Schultz (1950a) has pointed out that a mastoid process may occasionally be found in adult male gorillas, but it should be emphasized that, even where present, it differs markedly from the total morphological pattern that the hominid structure presents in the latter's sharply conical form, its well-defined posterior border (in apes the process has a flat posterior surface that represents a lateral extension of the nuchal area of the occiput), its flat medial surface bounded by a digastric fossa, and the details of its immediate relationship to the posterior margin of the auditory aperture. It also needs to be emphasized that, unlike that of the gorilla, the mastoid process appears to be a consistent (and not an occasional, sporadic) element of the australopithecine skull, and also that it is present even in the young, immature skull.

It is not possible to be certain of the functional significance of the hominid type of mastoid process in the australopithecine skull. Probably it is related to the relative degree of development and the direction of pull of the several muscles attached to it, and this is associated with the balance of the head on the spinal column. But it may also be a concomitant of the complex morphological changes that have affected the base of the skull and the relationships of its component elements.

e) *The Frontal and Upper Facial Regions.* The contour of the forehead and supraorbital eminences, combined with the orientation of the orbital apertures, the high level of the zygomatic arch, and the abbreviated temporal process of the zygomatic bone, make up (in Sterkfontein 5) a total morphological pattern that is surprisingly hominid and that does not appear to be paralleled in any pongid skull of equivalent size. It is even evident in the immature Taung skull. In Sterkfontein 1 the pattern appears also to have been closely similar (so far as can be determined from a reconstruction of this crushed specimen); but in the large Swartkrans and Olduvai skulls the frontal region is considerably flattened, as it also is in some of the skulls of *Homo erectus*.

f) *The Mandibular Fossa.* It has often been assumed that the type of movement that occurs at the mandibular joint can be determined by a study of the construction of the mandibular fossa of the skull. To a limited extent this is true, but the problem is complicated by the fact that in life the fossa and the condyle of the mandible are separated by an interposed fibrous disk of variable shape and thickness, which also needs to be taken into consideration because of the part it plays in the mechanics of the joint movements. The best indication of the nature of the jaw movements in the Australopithecinae is probably provided by the teeth and their mode of attrition. As we shall see, the dentition does give evidence of jaw movements similar to those of *Homo*, and it is particularly interesting, therefore, that the mandibular fossa also shows a combination of features of a hominid rather than a pongid type.

The mandibular fossa is especially well preserved in the Kromdraai skull, and in the details of its construction it conforms entirely with the hominid type. These details include a relatively deep articular concavity, bounded anteriorly by a prominent, transversely disposed, articular eminence; a postglenoid process of small size, which thus exposes the tympanic bone so that the latter forms practically the whole extent of the posterior wall of the fossa as a flattened, slightly concave, rectangular plate with a relatively sharp lower border; and no more than a slight indication of an entoglenoid process. In Sterkfontein 5 and 8 much the same construction is also found, except that the postglenoid process is more strongly developed (but a flattened tympanic plate still enters into the formation of the posterior wall of the mandibular fossa), and the same applies to the large Swartkrans and Olduvai skulls. Finally, the hollow of the mandibular fossa in all these specimens extends up to a consistently high level in relation to the auditory aperture, in some specimens reaching as high as the upper border of the latter (in the Pongidae the articular surface is almost invariably placed at a much lower level). All these constructional items of the mandibular fossa in the Australopithecinae (and particularly in the Kromdraai specimen) make up a total morphological pattern of a hominid type that has not been shown to occur in any of the known anthropoid apes. It is clearly a pattern of some intricacy, involving several different osseous elements, and may thus be regarded as of some importance for the morphological evidence it provides in assessing the affinities of the australopithecines.

Sutures

Certain authorities have laid considerable emphasis on variations of sutural pattern that they believe differentiate the Hominidae from the Pongidae, in particular the arrangement of the sutures in the pterionic region of the temporal fossa, the medial wall of the orbit, and the anterior fossa of the cranial cavity. In fact, however, these sutural patterns show some variation in each of the two families, and, so far as individual specimens are concerned, they cannot therefore be regarded as diagnostic. But it is of interest that in the australopithecine skull these patterns appear to conform to the normal hominid condition, that is, there is a sphenoparietal contact in the temporal fossa (reported to occur in at least three specimens), while an ethmolachrymal contact in the orbit and a sphenoethmoidal contact in the anterior fossa were probably present in Sterkfontein 5 (Broom, Robinson, and Schepers 1950). None of these sutural patterns is normally found in the African anthropoid apes (though they may all occur as occasional exceptions); on the other hand, they are normally present in the orang. Thus, although taken in combination they suggest a significant contrast with the African apes, they do not, of course, serve to differentiate the Australopithecinae from the Pongidae as a whole.

The high level of the temporoparietal suture on the cranial wall in the Australopithecinae is an indication of an expanded temporal squama, wherein these fossils show a rather obtrusive hominid character, for the relative size of this element is much smaller in the Pongidae than in the Hominidae. In anthropoid ape skulls we have examined the summit of the temporal squama was not found to extend above the horizontal level of the supraorbital margin in the Frankfurt plane. But this is a well-marked feature of Sterkfontein 5, in which the temporoparietal suture is very clearly seen.

Another suture that has been held to be of peculiar significance in connection with the relationships of the Hominidae is that related to the facial component of the premaxilla. This component is always found in the Pongidae (as in mammals generally) and in young individuals forms that part of the upper jaw in which the incisor teeth are implanted. In *H. sapiens* the premaxilla, as a separate element, does not normally enter into the formation of the facial surface of the upper jaw, for early in embryonic development it becomes fused with the maxilla. The suture has been reported in only two specimens of *Australopithecus*: Taung and Makapan 6. The reduction of the facial premaxilla may have been related to the recession of the jaws during quite a late stage in the hominid sequence of evolution, possibly even in the Early Pleistocene. The fact that the facial component of the premaxilla (with an extensive premaxillary suture extending up to the narial margin) has occasionally been observed in the skull of the newborn Negro clearly indicates that its genetic basis has by no means disappeared, even in *H. sapiens*.

Jaws and Palate

The jaws and palate constitute one of the most striking features of the skull of *Australopithecus*. The masticatory apparatus is always large in relation to the braincase (the neurocranium) and in some East African populations is very heavily developed indeed. In these creatures (usually classified as *Australopithecus boisei*) the proportions of the skull are striking, and, indeed, the size of the jaws accounts for the main differences between the different types of Australopithecinae. As the masticatory apparatus grows more and more powerful, all the bony structures concerned increase in thickness, and more area is necessarily available for the vast muscles that were evolved to operate the jaws. Not only are the areas of insertion very large on the skull and mandible, but the cross section of, for example, the temporal is very large and causes the zygomatic arches to flare widely. This development brought enormous forces to bear on the molar and to a lesser extent the premolar teeth. The australopithecine masticatory pattern will be discussed further below (under dentition).

The upper jaw and palate of the Australopithecinae show the marked reduction and recession of the incisor and canine region characteristic of *Homo*, with an evenly rounded alveolar margin. This is a very striking character of these creatures: both canine reduction and incisor reduction have in fact gone even further than in modern *H. sapiens*. (It appears that the rounded contour of the jaws is a simple product of this reduction.) Those populations with the most powerful masticatory apparatus seem to have carried this reduction of the front teeth to its greatest extent, while the molar series has been enlarged. The heavy jaws have a very robust horizontal ramus, nearly as broad as deep, and the symphyseal region is marked by both a superior and an inferior torus in many specimens, though the superior torus is more strongly developed. In modern apes the inferior torus or "simian shelf" is characteristic, while in man both tori have been replaced by the external buttress—the mental eminence, or chin.

The mental foramen is frequently single (in contrast to the multiple foramina characteristic of anthropoid apes), particularly in the Swartkrans specimens, and does not occupy the low position characteristic of the pongid mandible. As in the upper jaw, the contour of the alveolar border is definitely hominid in type and quite different from that of any known group of apes, living or extinct.

Apart from the hominid features of the australopithecine skull discussed above, a close study of some of the original specimens shows a number of other details that, though *individually* variable in both hominids and pongids, all together present a combination that still further emphasizes the hominid affinities of these South African fossils. Such details include the contour and construction of the supraorbital ridge, the relatively short and wide basiocciput, the angulation between the tympanic and petrous bones, the presence (usually) of a single infraorbital foramen, the inclination of the foramen magnum, the well-developed lingular process in the young mandible (a small process of bone overlapping the foramen on the inner aspect of the vertical ramus of the mandible), the disposition of some of the foramina in the cranial base, and so forth. Taken individually, as isolated abstractions, some of these features may not have a significant taxonomic relevance. But if they are all taken in combination with one another and with those features that are not to be found in any known group of pongids, they have a very high degree of taxonomic relevance, for they quite clearly compose a total morphological pattern of the hominid, not the pongid, type. On the basis of the skull structure alone, therefore, the proposition that the Australopithecinae actually represent an early phase in the hominid sequence of evolution appears to be well founded.

It may well be asked why this rather obvious interpretation of the skull structure was so vigorously contested when it was first put forward. One reason, as we have already pointed out, is that the earlier critics of Dart and Broom seem to have assumed that the absolute size of the brain by itself provides the final taxonomic criterion of distinction between the Pongidae and the Hominidae, which of course is by no means the case (see p. 141). Other features of the australopithecine skull that are mainly the secondary concomitants of a small braincase associated with large jaws, such as the marked prognathism and the development in some of the larger skulls of a small sagittal crest, were also taken to indicate pongid affinities, in spite of the fact that they must also be presumed to have been present in the earlier phases of the hominid sequence of evolution (i.e., before the rapid expansion of the brain to the size characteristic of *Homo* had begun to manifest itself). Mention has already been made of the sagittal crest in the Swartkrans and East African skulls (see p. 47), and here it is necessary only to reiterate that such a crest is not to be regarded as a morphological entity in itself with a separate genetic basis—it is the result of the upward growth on the side of the skull of temporal muscles that, in large adults, require an area of attachment more extensive than can be provided by the braincase itself. In any case, the crest (in those individuals in which it is developed) is very different from the sagittal crest that may be found in male gorillas, orangs, and (very occasionally) chimpanzees, in that it did not extend back into a high nuchal crest (see fig. 20). This point needs to be emphasized, for it had been argued (on the basis of one skull in which the occipital region is missing) that, since the crest must have been similar to that of a gorilla, the missing occiput must have shown a high nuchal crest; therefore the area for the nuchal musculature was very extensive; and therefore the head was held forward in a position incompatible with an erect posture. But the initial assumption that formed the starting point for this sequence of speculations has proved incorrect. For, without exception, in all the several occipital bones of the Australopithecinae that have now been discovered, the nuchal ridge is low and the nuchal area of limited extent, precisely as in hominid skulls (see figs. 21 and 22) and very different from gorilla skulls. This is the case with the

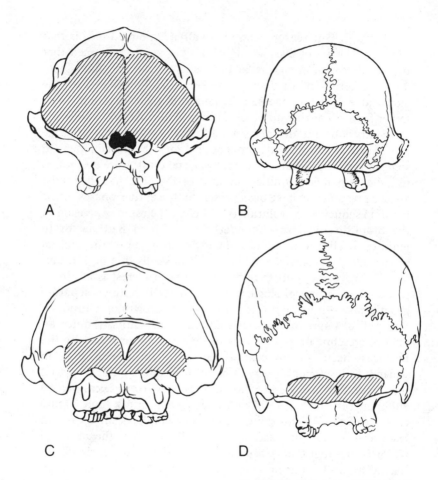

Fig. 22. Occipital view of the skull of *A*, a chimpanzee; *B, Australopithecus* (Sterkfontein 5); *C, Homo erectus* (after Weidenreich); *D, Homo sapiens* (European). Approximately one-third natural size. All the skulls have been oriented on the Frankfurt plane. Note the high level of the nuchal crest in the chimpanzee skull—a typical pongid character.

Olduvai skull and the Koobi Fora skulls, and it applies equally well to two skulls from Swartkrans with a small sagittal crest. Indeed, one of the specimens shows that the sagittal crest does not even extend back as far as the occipital bone. It needs to be recognized, of course, that the shelflike nuchal crest of the gorilla is produced not by the temporal muscle alone, but by the opposed growth of the temporal and nuchal muscles *together*. If, with the assumption of an erect posture, the nuchal musculature is reduced (as it is in the Hominidae) a shelflike nuchal crest is hardly to be expected.

Dentition

The dentition of the Australopithecinae has been studied in more detail than any other remains of these extinct creatures. There are several reasons for this. (1) Teeth have been found in great number and in many cases are excellently preserved. Apart from the permanent teeth, several examples of the deciduous dentition are also known. (2) The morphological characters of the dentition have proved of the greatest value for taxonomic determinations in the study of mammalian paleontology generally, and there is every reason to suppose that this applies as well to the Hominoidea as to other groups of mammals. (3) Because statements regarding the hominid features of the dentition were strongly contested by some critics, it became necessary to restudy all the teeth in greater detail to solve the conflict of opinion. It would be superfluous here to give a comprehensive account of the dental morphology of the Australopithecinae, for this information is available in readily accessible publications (Le Gros Clark 1950; Robinson 1956; Tobias 1967). That it conforms essentially to the hominid type may best be demonstrated by first listing and contrasting the fundamental characters that distinguish the Hominidae (as represented by *Homo*) from the Pongidae (table 4).

The differential characters of the dentition listed in table 4 are those that, on the basis of the comparative study of large numbers of Pleistocene hominids and pongids (both Recent and fossil), have been established to have a high degree of taxonomic

Table 4 Comparison of the Dentition of the Pongidae and *Homo*

Pongidae	Homo
Permanent Dentition	
The canine and postcanine teeth form approximately straight rows, parallel or (in fossil forms) slightly divergent anteriorly, and (with very rare exceptions) the upper dental arcade is interrupted by a diastemic interval related to the canines.	The dental arcade has an evenly curved contour of parabolic or elliptical form, with no diastemic interval (except to a slight degree in some individuals of *H. erectus*).
The incisors are hypertrophied and relatively broad (at least in the Recent genera of the large anthropoid apes). The sockets of the upper central incisors are placed well in front of the level of the anterior margin of the canine sockets (with occasional exceptions where the canines form unusually large tusks, e.g., in some large male gorillas).	The incisors avoid gross hypertrophy and thus retain more primitive proportions. The posterior margins of the sockets of the upper central incisors are on a level with, or behind, the anterior margins of the canine sockets.
The canines are relatively large, conical, and sharply pointed, with a well-marked internal cingulum commonly prolonged back (in the lower canine) into a talonid. The upper and lower canines overlap to a marked degree in occlusion; at an early stage of attrition they show facets on the anterior and posterior aspects of the crown and remain projecting well beyond the occlusal level of the postcanine teeth (except in very aged individuals, where all the teeth show an advanced state of extreme attrition or where the canine crowns may have been accidentally broken off during life). The canines show a pronounced sexual dimorphism.	The canines are relatively small, spatulate, and bluntly pointed, with the internal cingulum reduced to an inconspicuous basal tubercle. They wear down flat from the tip only and at an early stage of attrition do not project beyond the occlusal level of the postcanine teeth (except in some individuals of *H. erectus*). The canines show no pronounced sexual dimorphism.
The first upper premolar normally has three roots.	The first upper premolar commonly has one or two roots only, but in some modern races of *H. sapiens* a small percentage show three roots.

Pongidae	Homo
The anterior lower premolar is sectorial in character, commonly set obliquely to the axis of the tooth row, and with two roots. It is predominantly unicuspid, but it may have a rudimentary lingual cusp set on the lower slope of the large buccal cusp. In early stages of wear it shows an attrition facet on the anterolateral surface of the crown. There are no well-defined anterior and posterior foveae.	The anterior lower premolar is of a bicuspid type, with the cusps set in an approximately transverse plane. The two cusps are subequal in primitive hominids, such as *H. erectus*; but in *H. sapiens* the lingual cusp has undergone a secondary reduction. In primitive hominids the anterior and posterior foveae are sharply defined.
The molar teeth show considerable variation in their cusp pattern and in their absolute and relative dimensions. Except in advanced attrition (or rarely in abnormal specimens in which the canines have been accidentally broken off), the occlusal aspect does not become worn down to an even flat surface. In primitive pongids of Miocene and Pliocene age the first lower molar is relatively small.	The molar teeth show considerable variation in their cusp patterns and in their absolute and relative dimensions. In general, the cusps tend to be more rounded and more closely compacted than in the Pongidae. In *H. erectus* the second upper molar is larger than the first, and the last lower molar often greater in length than the second. In *H. sapiens* the second and third molars have undergone a secondary reduction. Even in early stages of attrition, the occlusal aspects of the molars commonly become worn down to an even, flat surface.
The canines erupt late, after the second molar and sometimes even after the third molar.	The canines erupt early, normally before the second molar (though a later eruption is stated to occur rarely in *H. erectus*, *H. neanderthalensis*, and in certain modern races of *H. sapiens*).

Deciduous Dentition

The lower canines are conical and sharply pointed, projecting well beyond the level of the milk molars, with an approximately straight internal cingulum extending back to form a talonid.	The lower canines are spatulate in form, not projecting markedly beyond the level of the milk molars, relatively short and bluntly pointed, and with no projecting talonid.

Pongidae	*Homo*
The first lower molars are sectorial in type, with the crown mainly composed of a single large, pointed cusp (the protoconid) and a depressed talonid. There is no well-marked anterior fovea.	The first lower molars are multicuspid, with four or five cusps disposed at approximately the same level. There is usually a well-marked anterior fovea.
The dental arcade is U-shaped.	The dental arcade forms an even parabolic or elliptical curve.

relevance. Taken together, they make up total morphological patterns of considerable complexity, which are distinctive of each of the two families. For the problem of the assessment of the affinities of the Australopithecinae, therefore, it is of particular importance to note that the total morphological pattern of the dentition of these fossil hominoids conforms to that of the later Hominidae (fig. 23). In all the adult specimens so far discovered,

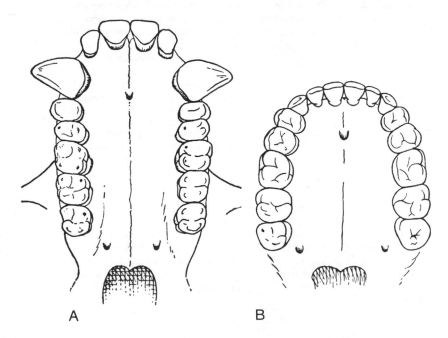

A   B

the dental arcade is evenly curved, with no diastemic intervals; the upper incisors are consistently small and retracted to the level of the canines; the latter are reduced in size and spatulate in form, show no very obtrusive sexual dimorphism, and, in the earliest stages of attrition, without exception became worn down flat from the tip to the level of the adjacent teeth; the anterior upper premolars have two roots (of which one may in some individuals be partially subdivided); the anterior lower premolars are of the nonsectorial bicuspid type (with the lingual almost as large as the buccal cusp) and have one root marked on the surface by longitudinal grooves that suggest a fusion of two roots; the molar teeth in the detailed morphology of their cusps show a much closer resemblance to the primitive hominid *H. erectus* than to any known type of pongid;[8] and in the earliest stages of attrition they became worn down to an even, flat surface. It is particularly important to recognize that these statements are based not on isolated and individual teeth but on a considerable number of specimens (in many cases with the whole upper or lower dentition complete or almost complete).

The same applies to the deciduous dentition, of which there are

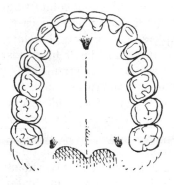

Fig. 23. The palate and upper dentition of *A*, a male gorilla; *B*, *Australopithecus*; and *C*, Australian aboriginal. Two-thirds natural size. The illustration of the australopithecine palate is a composite drawing made from several specimens, some of which are almost complete. (For illustrations of eight specimens of the australopithecine palate and upper dentition see Le Gros Clark 1950). Note in *Australopithecus* the relatively small canine and incisors, the absence of a diastema, and the evenly curved dental arcade. By courtesy of the Trustees of the British Museum.

C

at least four specimens (apart from isolated teeth). The milk canine is distinct from that of anthropoid apes, and the first milk molar is a complex tooth with four or five cusps at approximately the same level on the crown and a well-marked anterior fovea. The differentiation of australopithecine from ape canine teeth (both permanent and deciduous), which is so obviously apparent on direct visual comparison, has been fully demonstrated by the statistical analysis of appropriate measurements (Le Gros Clark 1950). The results of the multivariate analysis by Bronowski and Long (1952, 1953) are particularly revealing, for example, in their demonstration that, for all the milk canines so far discovered at that time, the discriminant functions they calculated fall well outside the range of modern anthropoid apes and actually lie within the range of *Homo*.

Finally, there is good evidence from the study of the dentition in immature specimens that the order of eruption of the permanent teeth conforms to that of the Hominidae. It may be noted that in the ascending scale of Primates there is a progressive acceleration in the replacement of the deciduous dentition relative to the eruption of the permanent molars, associated with a prolongation of the growth period. In the Pongidae the canine teeth are still very late in their replacement, often not completing their eruption until after that of the last molar and, in any case, not until the second molar has completed its eruption. The late eruption of the canine was retained in certain fossil hominids (e.g., *H. erectus*) and is still retained even in some groups of *H. sapiens*. But in most modern races the canine erupts before the second molar, and the first incisor tooth may occasionally erupt even before the first permanent molar. According to Schultz (1935), this occasional early eruption of the first incisor is a unique feature in which the Hominidae contrast with all other Primates. From the material collected at Swartkrans, it is evident that in this group of Australopithecinae the order of dental eruption actually corresponds to that normally found in *H. sapiens* (i.e., the canines erupted before the second molar), and in one specimen the first incisor has even erupted before the first permanent molar. On the other hand, the immature mandible found at Makapansgat shows that in this particular specimen the canine erupted after the second molar (as in *H. erectus* and some modern human races). The important point, however, is that the pattern of dental replacement characteristic of *H. sapiens*, and also present in some of the Australopithecinae, has not been found in any of the Pongidae, Recent or extinct.

It must be apparent from what has been written about the masticatory apparatus that the australopithecine dentition is not as a whole intermediate between that of modern man and modern ape, nor is it for the most part intermediate between that of a fossil ape, such as the Miocene *Dryopithecus*, and modern man. Although closely related to the human dentition and certainly hominid in form, it is characterized by a series of specializations that in some populations are of a most extreme kind.

The specializations are all related to a change in the masticatory adaptation that resulted, on one hand, in an enlargement of the molars and premolars (with "molarization" of the latter), and on the other hand, in a marked reduction of the canines and incisors. This generic adaptation reaches its greatest development in some of the very robust skulls and mandibles collected at Olduvai and Koobi Fora and believed to be between 2.0 and 1.0 million years of age. The molar teeth of these mandibles have extensive and complex folding of the enamel and thus have nearly twice the surface area of the molar teeth from Sterkfontein and thrice that of the molars of *Homo erectus*.

There is much variability, however, and in contrast to these immense East African specimens (such as Olduvai 5 and ER 729 from Koobi Fora) others are no larger than those of *Homo erectus* from Java (fig. 24). It is to be noted that the variability of modern man is also great, and that there is on the basis of these data no reason to suppose that very much reduction necessarily occurred in the molar series during hominid evolution until relatively recently, during the divergence of modern races.

It appears, on the other hand, that the large molarized populations of *Australopithecus* were extremely aberrant and highly specialized, so that they cannot be considered part of the evolving lineage leading to modern man. It is now clear that they

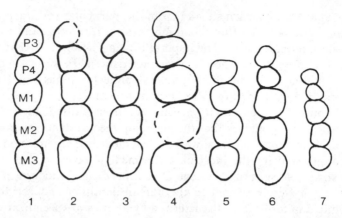

Fig. 24. Outlines (partly after von Koenigswald) of the lower molars and premolars of *1*, an orang; *2* and *3*, australopithecine specimens from Kromdraai and Sterkfontein; *4*, robust australopithecine mandible from Omo (Shungura); *5*, *H. erectus* (S1b); *6*, Australian aboriginal; *7*, European. Compare the difference between the overall dimensions of the australopithecine teeth and those of *H. erectus* with the differences which occur within the limits of the single species *Homo sapiens*. Approximately two-thirds natural size.

constituted a side branch of the lineage, of some importance in Late Pliocene and Early Pleistocene times (in Africa), that became extinct about 1.0 million years ago. These very robust populations are usually classified with the Olduvai 5 skull, which is the type specimen of *Australopithecus boisei*, and this species is now considered to constitute this specialized branch of the hominid lineage.

All the australopithecines, however, show some tendency toward this specialization, and the third lower molar (for example) commonly is longer than the second molar. The premolars are at least somewhat molariform, especially the fourth, and in the third the lingual cusp is large compared with that in modern man. The first milk molar is also highly molarized, with four or five cusps. The typical generic characters of *Australopithecus* also include (as we have mentioned) reduction in size of the canines and incisors, so that they appear disproportionately small in comparison with the large molar series. In those forms in which the molars have achieved such importance, the incisors and canines, especially in the mandible, are reduced to little more than pegs, and it is hard to see what purpose they could have served. While the incisors of modern *Homo sapiens* are often much reduced (as are his canines) they very rarely reach the small size seen in these heavy specimens of *Australopithecus*, and the modern human canine is usually a good deal larger.

The Lower Limb Skeleton

As we have already noted (p. 14), there is reason to suppose that the primary factor that determined the evolutionary segregation of the Hominidae from the Pongidae (probably in the Miocene or Early Pliocene) was the divergent modification of the limbs in adaptation to erect bipedalism. The evidence of *H. erectus*, for example, provides a positive indication that the lower limbs had become perfected for a fully erect posture and gait while the morphological features of the skull still retained a number of primitive characters approximating in some respects a simian level of development. If, then, the Australopithecinae are representatives of the hominid rather than the pongid sequence of evolution, the limb skeleton might be expected to be particularly relevant for the determination of their taxonomic status. In fact, this is the case. For example, several specimens of the australopithecine pelvic skeleton have been found, and they are consistent in their fundamentally hominid construction, as they are also in certain details in which they differ from the pelvis of modern *H. sapiens*.

The first pelvic bone was found in 1947 in the stalagmitic breccia of the Sterkfontein site, in direct association with fragments of a femur and tibia, some vertebrae, and a crushed australopithecine skull, and within a few feet of other skulls, cranial fragments and numerous teeth of the same creatures. Later, almost the whole of this pelvis was assembled, including the innominate bone of each side, the sacrum, and also (in an articulated condition) some of the associated thoracic and lumbar vertebrae (fig. 26). A small and fragmentary innominate was found at the same site in 1949. At Makapansgat, the ilium and part of the ischium of an adolescent was found in 1948, and some time later a second ilium was found (Dart 1949, 1958). In 1954, a fragmentary juvenile ilium was discovered at Kromdraai. A considerable part of a distorted australopithecine innominate was also discovered in 1950 at Swartkrans close to the remains of three skulls and associated with the lower half of a humerus. A second fragmentary innominate came to light in 1970. The Sterkfontein and Swartkrans specimens have been figured and briefly described by Broom and Robinson (1952) and Broom, Robinson, and Schepers (1950). After the 1947 Sterkfontein pelvis, probably the next most important postcranial discovery has come from Hadar, where in 1974 a skeleton 40 percent complete was collected by Dr. D. C. Johanson. It has much in common with the Sterkfontein specimen and includes the sacrum, left innominate, and left femur, associated with other bones. Detailed descriptions are, however, still awaited.

The hominid construction of the australopithecine pelvis shown by these specimens (figs. 25, 26, and 27) is marked by the relative breadth of the short ilium; the backward extension of the posterior extremity of the iliac crest and the low position of the sacral articulation in relation to the acetabulum; the orientation of the sacral articulation relative to the vertical axis of the innominate; the sharply angulated sciatic notch associated with a prominent ischial spine; the strongly developed anterior inferior iliac spine; the orientation and position of the ischial tuberosity in relation to the acetabulum (particularly in the first Makapansgat specimen); and the well-marked groove on the ventral surface of the ilium for the iliopsoas muscle. In all these characters, even individually, the pelvic bone makes a strong contrast with the modern anthropoid apes, and taken in combination they make up a total morphological pattern that is distinctive of the Hominidae among all other groups of mammals. Moreover, they are characters that are quite certainly related to posture. The broad ilium extends the anteroposterior attachment of the gluteal musculature (of the buttocks) that is used for balancing the trunk on the lower limbs; the bending-down of the posterior extremity of the

Fig. 25. Lateral view of three specimens of the australopithecine pelvic bone (5, from Sterkfontein; 6, from Makapansgat; 7, from Swartkrans) compared with those of *Homo sapiens* (4) and the modern large apes (1, gorilla; 2, chimpanzee; 3, orang). Approximately one-sixth natural size. The position of the ventral margin of the area of contact with the sacrum on the inner aspect of the bone has been indicated. Note that in its fundamental characters the australopithecine bone conforms to the hominid pattern. On the other hand, it contrasts strongly with the apes (this has been confirmed by comparison with eighty-seven pelvic bones of the modern large apes).

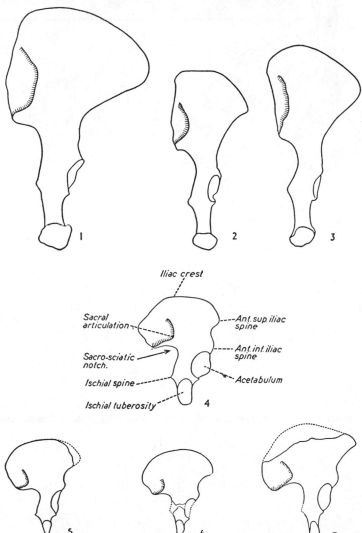

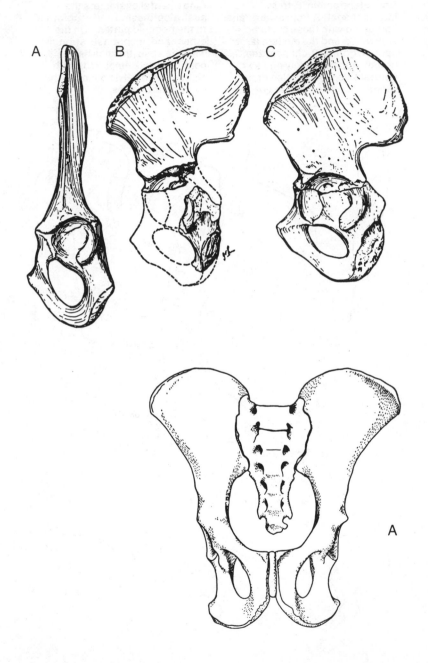

Fig. 26. Left lateral views of the innominate bones of *A*, an adolescent chimpanzee; *B*, an adolescent specimen of *Australopithecus* found at Makapansgat; and *C*, an adolescent Bushman. Note the outstanding hominid characters of the fossil specimen displayed by the relative breadth and orientation of the ilium, the sharply angulated sciatic notch, the strong development of the anterior inferior iliac spine, and the proximity of the ischial tuberosity to the acetabular socket. From Dart (1949).

iliac crest brings the gluteus maximus to a position behind (instead of lateral to) the hip joint and so permits it to play its essential role as an extensor in walking erect (see Washburn 1950); the approximation of the sacral articular surface to the acetabulum makes for greater stability in the transmission of the weight of the trunk to the hip joint; the reorientation of the sacral articulation is due to a rotation of the sacrum, which is associated with modifications in the disposition of the pelvic viscera; the angulated sciatic notch is a secondary result of the bending-down of the posterior part of the dorsum ilii and the accentuation of the ischial spine (which attaches a ligament strongly binding the pelvic bone to the sacrum); the robust anterior inferior spine

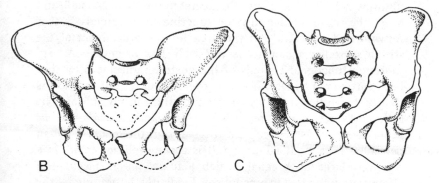

B   C

Fig. 27. The pelvis of *A*, a chimpanzee; *B, Australopithecus* (Sts 14); and *C*, a Bushman. All are drawn to the same scale from a photograph kindly supplied by Dr. J. T. Robinson. It is to be noted that, except for the lower half of the sacrum, the australopithecine pelvis is virtually complete, for those parts that were missing on one side of the fossil specimen have fortunately been preserved on the other side. From Clark (1971).

serves in part to attach the powerful iliofemoral ligament that braces the front of the hip joint in full extension in the standing position; the relatively high position of the ischial tuberosity enhances the extensor action of hamstring muscles in maintaining the position of full extension of the hip by bringing their upper attachments to a position behind, rather than below, the hip joint in the erect position (in anthropoid apes, and in pronograde mammals generally, the greater distance between the tuberosity and the hip joint permits greater power when the hamstrings are brought into use with the hind limb acting as a propulsive lever rather than a propulsive strut [Haxton 1947]); the deep groove for the iliopsoas muscle is related to the fact that in the habitually extended position of the thigh in the erect posture the muscle has to turn backward at a marked angle to reach its attachment to the femur. From considerations such as these it is reasonable to infer that the Australopithecinae had become adapted to an erect bipedalism, though there are indications that this was not exactly as in *H. sapiens*. For example, the ischial tuberosity (at least in the Sterkfontein and Swartkrans specimens) is not quite so closely approximated to the acetabulum as it is in *H. sapiens*, the acetabulum itself is small and deep, the region of the anterior superior spine extends farther forward and outward, and the area for the attachment of the strong sacroiliac ligaments is relatively less extensive. A very thorough functional analysis of the australopithecine pelvis has been carried out by Dr. Lovejoy (Lovejoy et al. 1973), and he has clearly shown that it is fully adapted to efficient bipedalism. The differences between the pelvises of *Homo* and *Australopithecus* listed above are associated with the fact that the modern human pelvis has been modified during Pleistocene evolution to admit a relatively large birth canal, which actually *lowers* mechanical efficiency with regard to bipedalism. Evidently the cranium of the newborn *Australopithecus*, with a brain only a little larger than that of an ape of similar size, presented no obstetric problems and permitted a lower interacetabular distance.

The functional implications of the australopithecine pelvic bone receive further corroboration from other parts of the lower-limb skeleton that have been found. We now possess four femora and a number of proximal and distal femoral fragments. The proximal femora (from Swartkrans, Olduvai, Koobi Fora, and Hadar), though clearly hominid, show certain characters that distinguish them from *H. sapiens*: they have a longer neck, a lower neck-shaft angle, and a smaller head. These characters are all consistent with the reduced interacetabular distance characteristic of the group. The distal femora (from Sterkfontein, Koobi Fora, and Hadar) also show a significant combination of characters, but in this region they are indistinguishable from those of *H. sapiens*. Of particular importance are the bicondylar angle and alignment of the condyles, the contour of the patellar surfaces with an elliptical lateral condyle with a high lip and deep patellar groove, and the forward projection of the intercondylar notch. The characters, taken together, make up a total morphological pattern that conforms to the modern human femur and unquestionably represents a mechanical adaptation to erect bipedalism. The disposition of the intercondylar notch by itself must be regarded as highly significant, for it is difficult to explain except on the assumption that (as in *Homo*) it accommodated the anterior cruciate ligament of the knee joint in the habitually extended position associated with an erect posture (Clark 1947).

The complete femur from the Hadar skeleton can be used to calculate an approximate stature of 3.5 to 4 feet. The poorly preserved specimen from Sterkfontein suggests a similar height for these gracile Australopithecinae.

The complete tibiae and proximal fragments known from Olduvai, Koobi Fora, and Hadar consistently show no characters that place them outside the range of modern man (in spite of claims to the contrary that were based on the Olduvai specimen). The path of the spiral line showing the tibial origin of the soleus muscle, the definition and form of the popliteal groove, the proximal joint surfaces and degree of platycnemia, the angles of torsion and inclination are all fully comparable to those characters in *H. sapiens*. The distal articulation with the talus is equally modern in form.

The Olduvai talus can be studied together with four specimens from Koobi Fora and an incomplete specimen from Kromdraai. The strong similarity between all these tali and modern tali has been recognized, but a number of peculiar features (such as the length and angle of the talar neck) have been stressed by different workers. Lovejoy, however, has made comparisons with a series of fifty modern Amerindian tali and finds that in no features do the fossil tali he has studied fall outside the range of variation of even such a limited sample of modern *H. sapiens* (Lovejoy 1978).

On the basis of these rather sparse data, no feature in the australopithecine foot is known to distinguish it from that of modern man. In fact, the whole of the australopithecine lower-limb skeleton, as we have seen, is similar to that of *H. sapiens* except in the single and relatively simple functional complex of characters related to the smaller interacetabular dimension. In practice this means that these creatures, as early as about 3.0 million years BP (the approximate age of the Hadar skeleton) were fully evolved bipeds.

Owing to the paucity of complete limb bones in the fossil record it has not until recently been possible to determine if the relatively long legs of *Homo sapiens* were present in the Australopithecinae. The Hadar skeleton has now enabled us to make some approximate estimations of limb proportions, and, as might have been predicted, this individual is reported to have hind legs significantly shorter (relative to arm length) than the mean of modern human populations. Further studies are needed to confirm this observation.

The Upper Limb Skeleton

The upper limb skeleton is less well represented in the fossil record than the lower limb. A small fragment of a scapula is known from Sterkfontein,[9] and a damaged clavicle shaft from Olduvai (perhaps belonging to *H. habilis*) together tell us little but are not certainly different from those of modern man. The humerus is known from Sterkfontein and Kromdraai (fragments), Koobi Fora (one complete robust specimen and five other fragments that await description), and Hadar (two gracile humeri of one individual as yet undescribed). There is also a gracile distal fragment from Kanapoi in Kenya, dated at 4.0 million years BP.

The analysis of these few poor specimens shows that the Kanapoi fragment was almost certainly very close in morphology to that of *H. sapiens*. The well-preserved Koobi Fora specimen, however, has some differences. In a multivariate study by McHenry and Corrucini (1975) this extremely large and robust bone appeared not to be allied to any group of existing primates, which might be taken to imply a functional uniqueness for at least one population of robust australopithecines.

Fragments of radii are known from Makapansgat, Olduvai, and Koobi Fora, as well as from the Hadar skeleton. None of those described are said to be distinct from that of modern man. There is a proximal ulna fragment from Kromdraai and an almost complete ulna from Omo, as well as two from the Hadar skeleton. That from Kromdraai is described as similar to that of *H. sapiens*, while that from Omo is said to share some arboreal adaptations with the Koobi Fora humerus described above. Though it is as long, however, it lacks the powerful muscular markings seen in the latter. Further comparative work here would be of value.

A capitate (wrist bone) from Sterkfontein and three metacarpals from Swartkrans are known. The capitate, associated with *A. africanus*, has characters intermediate between ape and man (Le Gros Clark 1967). Two of the metacarpals appear indistinguishable from those of modern man and should be classified with the *Homo* remains from this site, while the third (SK 84)—a thumb bone—has been attributed to *A. robustus* on the grounds of its size and great robusticity. Its morphology suggests functional capability of an extremely powerful grip (Napier 1959). Hand and foot bones from Koobi Fora and Hadar still await detailed study.

The elements of the upper-limb skeleton so far discovered and described are consistent with the inference of erect bipedalism; for, again taken in combination, they conform closely in many details to the corresponding parts of the upper-limb skeleton of *Homo* and differ in several respects from those of the powerfully developed limbs of Recent anthropoid apes (Clark 1947). They certainly do not show the pattern of characters that may be regarded as distinctive of upper limbs used either for quadrupedal progression or for a specialized type of brachiation.

So far as the posture is concerned, it will be observed that the evidence of the pelvis is confirmed by the lower limb bones, and the evidence of the limb skeleton all together is confirmed by morphological features of the skull. In other words, all the anatomical evidence is consistent within itself. That is to say, it indicates an erect posture and gait manlike, not apelike, in character. It is interesting that it is also consistent with the climatological evidence, for, as already pointed out, this indicates that the Australopithecinae were able to live in comparatively arid areas. It seems certain, therefore, that they must have been well adapted to life in open terrain of the sort that still exists in the savannas of today.

The Taxonomic Status of the Australopithecinae

It is now clear that, as between the Pongidae and Hominidae, the Australopithecinae must be allocated to the latter family. They show none of the divergent modifications that are distinctive of the Pongidae, their only resemblances to the latter being the retention in common with them of primitive hominoid characters, such as the small size of the brain, large molar teeth, and a few very minor features of the limb skeleton. As was already discussed in detail, such primitive features presumably must have been present in the earlier phases of hominid evolution (as, indeed, they still were to some extent in *H. erectus*). The only other possible interpretation is that the Australopithecinae represent a third (and hitherto unknown) radiation of the Hominoidea with no particular relationship to either the Hominidae or the Pongidae, but showing a most extraordinary parallelism with the former. But such an interpretation would be wholly gratuitous, with no valid supporting evidence, and it would demand a degree of evolutionary parallelism far beyond anything that has been demonstrated to have occurred in any other mammalian sequence of evolution.[10]

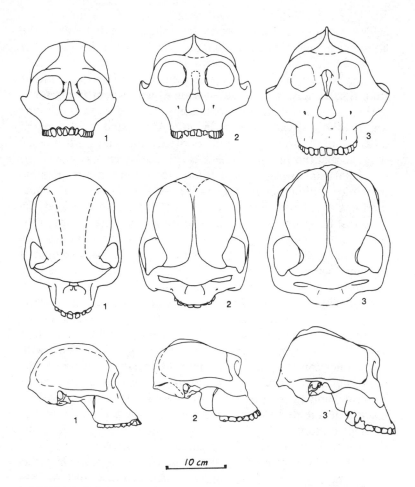

10 cm

Fig. 28. Frontal (*above*), superior (*middle*) and lateral (*below*) views of *Australopithecus africanus* (*1*) *A. robustus* (*2*), and *A. boisei* (*3*). Note especially the cranial characteristics that change with increasing cheek-tooth size (*A. africanus* having the smallest teeth, *A. boisei* the largest).

Considered as a series, the face is flattest and deepest, and bony buttresses are better developed, and the temporal fossa (indicating temporal muscle bulk) is largest, in *A. boisei*. All these features are related to the size and mechanical efficiency of the posterior dentition. After Tobias 1967.

The recognition that the Australopithecinae belong to the hominid sequence of evolution does not necessarily mean, of course, that they were ancestral to *Homo*. Certainly, this is unlikely to have been the case with some of the later populations from Swartkrans, Kromdraai, Olduvai, and Koobi Fora; for it now seems probable that these existed at a time when *H. erectus* was probably already present in both Africa and Asia. On the other hand, the greater antiquity of the Laetolil and Hadar fossils suggests that the australopithecines of East Africa may have been the evolutionary precursors of *Homo*, though it remains possible that some still earlier representatives of the Australopithecinae (perhaps in another part of the world) may have provided the ancestral basis from which all later hominids were derived. On purely morphological grounds there seems to be no serious objection to this interpretation, for in their skeletal and dental anatomy some of the Australopithecinae conform very closely indeed to theoretical postulates for a phase of hominid evolution immediately preceding *Homo*. It is true that certain of their morphological features, such as the relatively large size of the premolars and molars and the degree of molarization of the first deciduous molar, have been interpreted as specializations, with the implication that, not being present in later hominids, they could not have been present in the ancestors of the latter. However, only in the robust populations do these specializations reach a degree that would place them outside a possible lineage leading to *Homo*. These robust populations also happen to be those that lived too late to be human ancestors.

We have been using the term "Australopithecinae" in reference to the whole collection of South African and East African fossils as a temporary device, to avoid confusion while the taxonomic nomenclature of the various groups still remains a matter of disagreement. And, as previously noted, the view taken here is that such anatomical differences as there may be between the fossil remains found at the different sites in the Transvaal, East Africa and northeast Africa, though they may possibly justify subspecific, or even specific, distinctions, are not sufficient to justify generic distinctions. That is to say, the differences do not appear to equate in degree with those that are usually regarded as an adequate basis for generic distinctions in other groups of the Hominoidea. But the problem of the taxonomy of the Australo-pithecinae is basically a matter of the definition of terms; and until this has been attempted, it is hardly possible to achieve final agreement. On the assumption that all the australopithecine remains so far discovered represent a single genus, *Australopithecus*, this genus may be provisionally defined as follows.

Known cranial capacity ranging from ca. 420–530 cc; relatively thin-walled cranium with strongly built ectocranial superstructures and medium to robust supraorbital torus; heavily pneumatized cranial base; endocranial cast suggests posterior position of lunate sulcus; in populations with very heavy masticatory apparatus a low sagittal crest in the frontoparietal region, widely flared zygomatic arches, large temporal fossae, zygoma arising from maxilla between C and M1; large and prognathic face; occipital condyles behind the midpoint of the cranial length, and on a transverse level with the auditory meatus; nuchal area of the occipital restricted and downward facing as in *Homo*; consistent development of pyramidal mastoid process; mandibular fossa on pattern of *Homo* but sometimes with pronounced postglenoid process; jaws variable in size but often massive, especially in mandibular corpus; vertical ascending ramus; chin absent, but symphyseal region relatively vertical, with inner surface strengthened by superior and inferior transverse tori.

Dental arcade parabolic in form with no diastema; incisors relatively reduced, especially in robust populations; moderate-sized spatulate canines wearing down from the tip, showing only slight dimorphism; anterior lower premolar bicuspid or multicuspid with subequal cusps; premolars relatively large especially in buccolingual diameter and molars often very large with multiple cusps, with progressive increase in size of permanent lower molars from M1 to M3; consistent molarization of first deciduous molar.

So far as is known the postcranial skeleton conforms in its main features to that of modern man except: the upper limb of robust populations relatively long and muscular, possibly adapted to arboreal climbing, the hand very powerful; in the pelvis the interacetabular distance and birth canal relatively small, iliac blades widely flaring, small sacroiliac articulation and acetabulum; small femoral head and long angled neck.

This diagnosis includes all the genera and species listed on page 135.

The question still remains whether this material justifies a subdivision of the genus in terms of species or subspecies. As we have seen, there are certain morphological differences between the fossil individuals found at the different sites, and these permit a broad subdivision with two main groups. Some of the material found at Olduvai, Omo, and Koobi Fora is exceptionally robust, with massive masticatory apparatus, and has been associated with less robust specimens from Kromdraai and Swartkrans in South Africa. Robinson (1954), even on the basis of the South African sites alone, maintains that the two groups are generically distinct, and, because of his intimate acquaintance with most of the orginal material, his opinion deserves close consideration. But it has also been suggested that they bear a relation to one another that is comparable to that between the pygmy and other races of modern *H. sapiens* or between the common and pygmy chimpanzee. If this latter analogy is strictly appropriate, there can hardly be any justification for according them even a specific distinction. Oakley advocates recognizing two species of *Australopithecus*, an earlier *A. africanus* and a later *A. robustus*, on the basis of the South African material; most writers now recognize the distinction of the East African superrobust specimens and classify them as *A. boisei*. As a provisional taxonomic device for distinguishing the main groups, this is perhaps a reasonable compromise. Of these types the former was smaller, lighter in build, and more slender in skeletal structure; the latter was a larger creature, more heavily built with a coarser skeletal structure, and showed a pronounced development of the masticatory apparatus. On the basis of such differences it has been suggested that *A. africanus* is a more "advanced" type than *A. robustus*, but it seems more reasonable to regard it as a more generalized type of the same genus and thus as more likely to have provided the ancestral basis for the subsequent evolution of *Homo*. If this interpretation is correct, it is to be assumed that *A. robustus* and *A. boisei* were specialized types, somewhat divergent from the main line of hominid evolution. Further, if there were more or less contemporaneous and already distinct species in the African continent during early Pleistocene times, it follows that the divergence of the two types must have occurred earlier, in the Pliocene.

As we have seen (chap. 3) a number of specimens from Olduvai Gorge dated 1.8–1.6 million years BP have been made the types of a third species of *Homo*: *Homo habilis*. Some of these are very fragmentary, and their morphology is not as well understood as that of other groups discussed in this chapter. In general they have much in common with *Australopithecus africanus* from Sterkfontein and with *Homo erectus* fossils from Sangiran and have been considered by some authors either very late *Australopithecus* or early specimens of *Homo erectus*.

A few specimens were collected from Swartkrans in 1949 that were recognizably distinct from the rest and were derived from a pocket of breccia that is now known to have been deposited at a later date than the main mass of bone-bearing breccia. These specimens were named *Telanthropus capensis* at their discovery, but, as we have seen, they and a fragmentary cranium found in 1952 are now considered the South African equivalent of the intermediate *Homo habilis* in East Africa and are of about the same age. (The mandible SK 15 is better classified as *Homo erectus*.)

From Koobi Fora we have specimens from the Lower Koobi Fora Formation that seem to fall into this intermediate category and are described by Richard Leakey simply as *Homo*. Most authorities now recognize the intermediate nature of all these specimens, but some still question whether the creation of a new species is justified. It might be predicted that specimens that lie on the boundary between two sequent genera (or species) would prove controversial in interpretation. Further analysis of all these specimens (especially those from Koobi Fora) is needed before a justifiable taxonomic assessment can finally be made (see p. 123).

Little is known of the activities or mode of life of *Australopithecus*. A number of stone implements have been found at two sites alongside *Australopithecus* remains (Swartkrans and Olduvai bed I), but in both cases there is recognized evidence for the presence of a more advanced hominid that is classified as *Homo habilis*. There is at present no reason to attribute tool-making to either *Australopithecus robustus* or *A. boisei*,[11] and no sites contain a recognizable industry that predates that at Omo-Shungura (member F), reliably dated at ca. 2.0 million years BP.[12] As we have indicated, this suggests that recognizable tools first appear with the coming of *Homo*, though sporadic tool-making must have begun at an earlier date, especially in materials that are easier to work than stone. Attempts have been made to credit these creatures with the ability to use bone implements and even with the ability to make fire, but the evidence for the latter assumption has proved to be faulty. An indication that the Australopithecinae may have used weapons of some sort is provided by a large number of baboon (*Parapapio*) skulls found associated with their remains, for (according to Barbour 1949), out of fifty-eight of these skulls, no fewer than forty-two (72 percent) show evidence of depressed fractures in the parietal region that are rather consistent in position and extent. It has been suggested that these injuries must have been the result of well-aimed blows with an implement of some sort. However, postmortem damage and distortion of skulls during fossilization is extremely common, and while this hypothesis is by no means disproved, it should be treated with caution. Nevertheless, there are obviously good reasons to suppose that the Australopithecinae were endowed with an intelligence and skill superior to that of the modern anthropoid apes which would have been effective in the development of predatory behavior.

General Observations on the Australopithecinae

The australopithecine fossils have been discussed in some detail because of their great importance for the study of human phylogeny. The remarkable quantity of the material so far discovered has already provided an unusual amount of information regarding the anatomical structure of these extinct creatures, but its very abundance means that its detailed analysis can be completed only after many years of study. It is clear, however, that the total morphological pattern presented by the skull, teeth, and postcranial skeleton conforms to that of the Hominidae rather than the Pongidae, in spite of the small size of the brain. But, though they are certainly hominids in the taxonomic sense, the terms "man" and "human" can be applied to them only with some reserve, for there is no certain evidence that they possessed any of the special attributes commonly associated with the human beings of today. They are to be regarded, rather, as representatives of the prehuman phase of the hominid sequence of evolution. In their morphology they appear to conform very closely to theoretical postulates for the immediate evolutionary precursors of the *Homo* phase of hominid evolution; and it is for this reason that the genus *Australopithecus* has been provisionally suggested as the ancestral stock (or at least very closely related to the ancestral stock) from which more advanced species of the Hominidae were derived. However, this interpretation depends on the demonstration that the genus actually does fit into the postulated *temporal* sequence. We have noted the evidence that some of the fossils date from the Pliocene period, and it has been estimated, on the basis of potassium-argon datings, that australopithecines inhabited East Africa well over three million years ago (at Laetolil). Thus there is no theoretical difficulty (from the point of view of the geological sequence) in accepting *Australopithecus* as ancestral to *Homo*. Even so, however, it would not necessarily follow that the transition occurred in Africa. It may have occurred in some other part of the world, and the African fossils in that case may represent but slightly modified survivors of the ancestral stock.

Obviously, the precise position of *Australopithecus* in hominid phylogeny can be determined only by a more complete paleontological record. But it may be emphasized that, even if evidence of geological antiquity were totally lacking, the purely morphological evidence of the Australopithecinae would still be highly significant. For they demonstrate (as had already been predicated from a consideration of comparative anatomical data and of the paleoanthropological sequence leading back to *H. erectus* and *H. habilis*) that there once existed primitive hominids with a cranial capacity exceeding by very little that of the large anthropoid apes but with a limb structure evidently related to the development of an erect posture and gait that is so marked a characteristic of the evolutionary sequence of the Hominidae in general.

Five

The Origin of
the Hominidae

Undoubtedly the most intriguing question in the whole evolutionary story is, What was the ultimate origin of man? Or, put in zoological phraseology, When in geological time did the Hominidae become finally segregated from other groups of the Primates, and what was the nature of the ancestral stock from which this segregation occurred? Unfortunately, any answers that can at present be given to these questions are based to some extent on indirect evidence and thus are somewhat conjectural, for the paleontological record of the earliest Hominidae is still incomplete. The gradational series of types—modern *Homo sapiens*, Early *Homo sapiens*, *Homo erectus*, *Homo habilis*, and *Australopithecus*—compose a retrospective sequence morphologically, and also a receding temporal sequence (fig. 29). They thus seem to provide satisfactory evidence for carrying our ancestral lineage back to a phase of hominid evolution to which the term "human" can no longer be applied—when the size of the brain was little greater than that of the modern anthropoid apes and the jaws were massive and protruding. But it is a phase that still appears (in spite of the small brain) to have been well advanced beyond the initial origin of the Hominidae, that is to say, long after the time of their evolutionary divergence from the Pongidae. For example, many characteristic features of the hominid skull and dentition had already been established by the *Australopithecus* phase, and

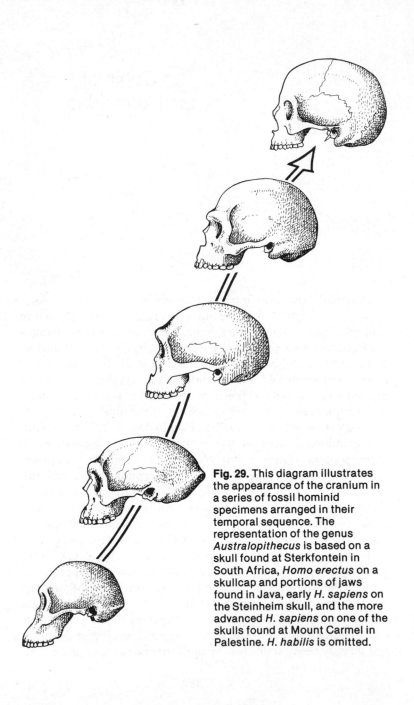

Fig. 29. This diagram illustrates the appearance of the cranium in a series of fossil hominid specimens arranged in their temporal sequence. The representation of the genus *Australopithecus* is based on a skull found at Sterkfontein in South Africa, *Homo erectus* on a skullcap and portions of jaws found in Java, early *H. sapiens* on the Steinheim skull, and the more advanced *H. sapiens* on one of the skulls found at Mount Carmel in Palestine. *H. habilis* is omitted.

(what is still more important) the limb skeleton had already undergone modification in adaptation to an erect posture and gait, very different indeed from that which must be postulated, on the basis of comparative anatomical evidence, for a common ancestral stock. The fact is that the most serious hiatus now in the record of hominid evolution is the gap that separates the genus *Australopithecus* from the fossil hominoids of Miocene times. It is true that, by extrapolation backward and by analogy with what is known of the paleontological history of other mammalian groups, we can contrive a theoretical picture of the intermediate stages that presumably must have been interposed between generalized pongid ancestors and the *Australopithecus* phase; but this is not a very satisfying procedure.

More than twenty different genera of large extinct hominoids of Miocene age have been described from sites in Europe, Asia, and Africa, but critical studies in recent years indicate fairly clearly that these fossils represent at most only five distinct genera,[1] that is, *Dryopithecus, Proconsul, Sivapithecus, Gigantopithecus*, and *Ramapithecus*, and many consider *Proconsul* to be a subgenus of *Dryopithecus* (Simons and Pilbeam 1965). Species of *Dryopithecus* apes are known from Africa throughout the Miocene, and from about 16 million years BP (after the African continental plate touched Eurasia and a land bridge was established) the genus spread rapidly throughout the then warm forested regions of Eurasia from southwest France across to China. Most of the fossil remains are gnathic, though a few limb bones are known. The best-preserved skull is of *Dryopithecus (Proconsul) africanus*, from Rusinga Island in Kenya, and carries a date of ca. 18 million years BP. As may be seen in figure 30, this skull is of a generalized type, with no more than a moderate prognathism and without the supraorbital torus and high nuchal crest of the modern African apes. It is to be noted, however, that this is an isolated example and may not therefore be assumed to be typical of Miocene apes in general. Miocene hominoid jaws and teeth are numerous but very fragmentary. In the majority of specimens the dentition conforms in its general features to the pongid type, with projecting and conical canines, sectorial lower

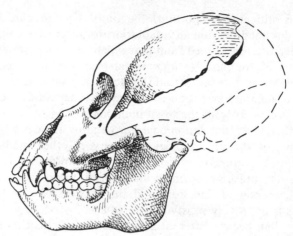

Fig. 30. The skull of the Miocene ape, *Proconsul africanus*, as reconstructed by Davis and Napier (1963).

first premolars, and elongated molars that are usually large and, in the lower dentition, increase progressively in size from front to back. On the other hand, the cusp pattern of the molars can be distinguished from that of the modern large apes and is in general less complicated: the incisor teeth are relatively smaller (and in this respect more closely resemble hominid incisors), and the simian shelf of the mandible is either absent or developed only in an incipient form. In other words, the dentition and mandible of these early hominoids were less specialized than in Recent Pongidae.

Of special importance is the Miocene hominoid *Ramapithecus*, first described by Gregory, Hellman, and Lewis (1938) from the Siwalik deposits of India, which, in the small size of the teeth and the relatively simple construction of the molars, seems to approximate more closely the Hominidae than do the other hominoid genera (but see Frayer 1974*b*).

In the past fifteen years more specimens of this interesting Primate have been identified in museum collections from the Siwalik Hills, and further discoveries have been made in this region by von Koenigswald and Pilbeam. All together, some fifteen specimens of mandibular and maxillary fragments from this area are now known from the formations named Chinji, Nagri, and Dhok Pathan (see fig. 31). These fossiliferous deposits extend through the Late Miocene into the base of the Pliocene. The fauna accompanying the fossils indicates a rain forest (Chinji) giving way gradually to a grassland steppe environment (Dhok Pathan). This is precisely the kind of environmental situation in which it has so often been postulated that the earliest hominid evolution would have occurred.

The specimens collected and identified as *Ramapithecus* have the following characteristics that relate in a very precise way to the particular masticatory adaptations of the group. The molar teeth are low-crowned and steep-sided, with broad, flat occlusal surfaces. They are tightly packed, have thick enamel, and show a high rate of wear. The premolars are bicuspid (but the lower first premolar has a small if distinct lingual cusp).

Only the canine socket is known, but compared with *Dryopithecus* or the living apes it is relatively small: the orientation of the long axis of the socket, its medial position in the lower dentition, the absence of a pronounced diastema, and the low degree of prognathism are all suggestive of the hominid pattern. The incisors (judging by their sockets) were extremely small and nearly reach the very reduced condition seen in *Australopithecus*. In the maxilla the zygomatic process can be seen to originate above the first molar tooth as in the hominids (rather than above the second or third), and it is probably somewhat flared. This suggests large and forward-placed temporal and masseter muscles, the main sources of power for chewing and grinding food. This trait, together with the deeply arched palate, the short and deep face, and molars of the kind described above, is highly suggestive of an adaptation to a tough diet requiring powerful mastication.

These characters are very suggestive of the sort of transitional phase that may be presumed to have occurred in the evolutionary derivation of the hominid type of dentition from that characteristic of known Miocene apes. Indeed, Simons (1961) follows Lewis

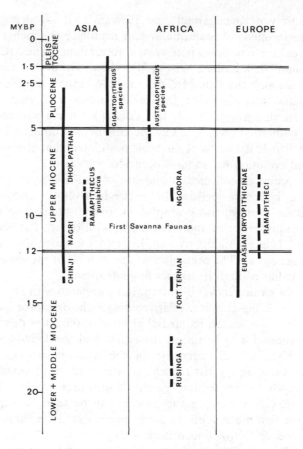

Fig. 31. Ranges of uncertainty in
the ages of important Pliocene
and Miocene deposits and
Hominid and Pongid fossils.

(in Gregory, Hellman, and Lewis 1938) and inclines to the view
that the total morphological pattern provided by these features in
combination indicates that the genus should be more properly
allocated to the Hominidae than to the Pongidae. The populations that these fossil fragments represent, and that span such a

long period of geological time, survived in the more open grassland conditions long after most of the *Dryopithecus* apes, which had been their contemporaries, had died out. However, the relevant fossil material from India is still too scanty to permit any definite formulation of an evolutionary sequence adumbrating the initial divergence of the Hominidae, even though the indications provided by the few Siwalik specimens already obtained are sufficiently interesting to arouse the hope that this region will sometime yield remains of the rest of the skull and skeleton. Only these will indicate for sure if these fragmentary remains are in fact those of the earliest Hominidae. The region certainly does seem to offer rich prospects for the paleoanthropologist.

Further remains that can be attributed to *Ramapithecus* were discovered in 1961–62 by Louis Leakey at Fort Ternan, Kenya, in deposits older than those of India and Pakistan and dated radiometrically to about 14 million years BP (fig. 31, Walker and Andrews 1973). The remains consist of the maxillary and mandibular fragments of one individual and have been reconstructed to indicate the form of the palate and mandible (figs. 32, 33). This is clearly a far more apelike dentition than that described above, with larger canines, a diastema, and a longer, narrower palate. However, there are hominid features in the simple molar teeth, the broad flat face, and the small incisors. Above all, we find here a lower first premolar with an incipient lingual cusp that, together with the relatively reduced canines and short face, separates the mandible from those of the contemporary fossil *Dryopitheci*.

Leakey named this specimen *Kenyapithecus wickeri*. Since, however, it has features in common with the Indian *Ramapithecus*, and since a list of diagnostic characters sufficiently distinct to justify the creation of this genus has not been published, it is usual to classify this specimen as *Ramapithecus wickeri* until further evidence is available (Simons 1963). The earlier age of the specimen, compared with the Indian fossils, does fit its slightly more apelike morphology. It seems possible that it could represent a population of hominoid Primates that had evolved some hominidlike adaptations in their dentition.

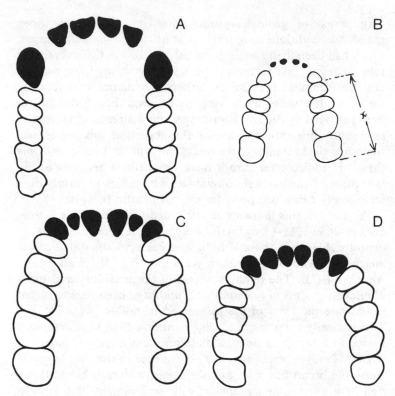

Fig. 32. Palatal view in outline of the upper dentition of *A*, a chimpanzee; *B, Ramapithecus*; *C, Australopithecus*; and *D*, an Australian aboriginal. Approximately natural size. In all cases the incisors and canines are represented by their sockets only. The sockets in *Ramapithecus* have been in part reconstructed.

Since these discoveries in India, Pakistan, and Kenya, further finds of *Ramapithecus*like Primates have been reported from Hungary, Greece, and Turkey. These vary in age and ecological association between those from Hungary (Rudabanya), and Turkey (Pasalar) which are probably of Middle Miocene age and are accompanied by a forest fauna (Kretzoi 1975; Andrews and Tobien 1977), and that from Greece (Athens), which is of Upper Miocene age and is accompanied by a grassland fauna.

Owing to the fragmentary nature of all these finds and a lack of detailed descriptive publications, it is not yet possible to work out relationships or a satisfactory taxonomy. In general, however, it is clear that populations of apes in different areas of Eurasia, and possibly in East Africa as well, were becoming adapted to the changing ecological conditions developing during the Upper Miocene. As the forest-living *Dryopitheci* became extinct together with the forest they inhabited, the *Ramapitheci* in different places adapted to the changing conditions and survived the disappearance of the forest. Adaptation to a grassland environment

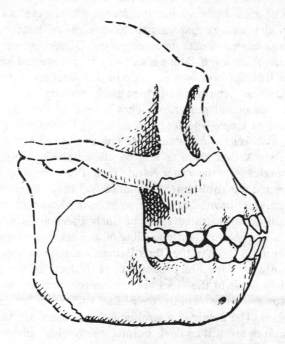

Fig. 33. *Ramapithecus punjabicus.* Lateral view of reconstructed jaws (after Simons 1977). Approximately one-half natural size.

certainly seems to have characterized the hominid lineage, but whether these populations of *Ramapithecus* adopted the hominid mode of bipedalism in their adaptation (and thus can be truly termed Hominidae) cannot yet be determined, until postcranial remains are discovered and the existence of bipedal adaptations demonstrated. Since such adaptations are well advanced in the Pliocene *Australopithecus*, we know that they must have evolved during Upper Miocene and Early Pliocene times. Just where these Hominid adaptations first appeared we do not know, but *Ramapithecus* is known from both Africa and Asia by 14 million years BP, and the ancestral Dryopithecine apes were also present in both continents (Campbell and Bernor 1976).

It is of great interest that the *Ramapitheci* were not the only hominoid primates that carried clear signs of adaptation to a grassland environment. Another genus, *Gigantopithecus*, shows a parallel adaptation. This genus was first recognized and named by von Koenigswald on the basis of a collection of fossil teeth from Chinese pharmacies (where such material was commonly sold for medicinal purposes). They were described by von Koenigswald as a new species and genus of Pleistocene fossil ape, *Gigantopithecus blacki*, which he saw as the last survivors of an Asiatic stock that more or less paralleled the hominid lineage. Weidenreich, on the other hand, took the view that the teeth represented a giant hominid (1946). Since then, three mandibles of the creature have been recovered from the Kwangsi Province of China in deposits stated to be of Early Pleistocene age (Chang 1962), and another giant mandible of a much earlier species has been found in India in the Dhok Pathan zone of the Siwalik Hills.

Broadly speaking, the proportions of the teeth of this genus are similar to those of the Hominidae, and the canines, which have low crowns, wear rapidly to a flat surface to resemble the premolars. The hominid resemblances, however, are superficial. The canines are still conical, pointed teeth when unworn and do not resemble the hominid incisiform canine, and the mandibles are very heavy and apelike. But they do exhibit broadly the same adaptive complex associated with a powerful masticatory apparatus, which can process very tough vegetation, and they are associated with a grassland, steppe fauna. These fossils can easily be derived from the earlier Siwalik *Dryopithecus indicus*, so that it is possible to postulate the fossil lineage: *D. indicus, G. bilaspurensis, G. blacki*. Although *Gigantopithecus* is still claimed as a possible hominid ancestor by some writers (e.g., Eckhart 1975), it seems clear that it lived too late to be ancestral to either *Australopithecus* or *Ramapithecus*, and it carries fewer hominid features than does *Ramapithecus*. Thus the best interpretation that can be placed on the present evidence is that *Gigantopithecus* is a (most instructive) parallel hominoid adaptation to a broadly similar environment (see Frayer 1974*a*).

The fossil genus *Sivapithecus* was sunk into the genus *Dryopithecus* in the important review by Simons and Pilbeam (1965); the so-called generic characters of the genus did not hold up under careful comparison with the closely related *Dryopithecus*. Recent discoveries in Eurasia and further analysis of other finds, however, have presented evidence that *Sivapithecus* is indeed probably distinct at the generic level from *Dryopithecus* and, like *Gigantopithecus*, carries dental adaptations associated with a grassland environment. Pilbeam reports thick enamel and relatively large molar teeth such as we find in the other grassland apes and in *Ramapithecus*, but in *Sivapithecus* the canines remain large as in the modern Pongidae. All this evidence suggests that the Hominidae were only part of the hominoid grassland radiation that appeared toward the end of the Miocene. They were, however, the only such lineage to survive the Pleistocene.

Reference should also be made here to the Miocene ape *Oreopithecus*, of which jaws with the dentition were first described almost a hundred years ago. In more recent times much richer material of this fossil hominoid has been found in lignite deposits in Tuscany by Hürzeler (1958), including an almost complete (but badly crushed) skeleton. In a number of characters *Oreopithecus* does show an approximation to the Hominidae, for example, in the small size of the canine tooth, the bicuspid character of the front lower premolar, the vertical symphysis of the mandible, and the breadth of the iliac bone. On the other hand, the lower molar teeth display certain cercopithecoid traits (particularly in the elongated talonid of the third molar), the increased breadth of the iliac bone involves its anterior region (and not, as in the hominid pelvis, the posterior region), and the relative elongation of the upper limbs clearly indicates an advanced degree of specialized brachiation very similar to that of the modern anthropoid apes. There is now general agreement that while *Oreopithecus* is certainly a hominoid (and not, as at one time supposed, a deviant type of cercopithecoid), the genus must be regarded taxonomically as a somewhat aberrant pongid retaining a few residual cercopithecoid characters. Hürzeler's suggestion that it is a primitive hominid having a place in the direct ancestry of man has not been found acceptable.

The question now arises whether there is any theoretical objection to postulating the derivation of the hominid type of dentition from the pongid type of dentition represented in the known Miocene apes of generalized type such as *Dryopithecus*. It has indeed been argued that the projecting conical canines and the sectorialized lower first premolars found in these apes (wherein they approximate closely the modern apes) are specializations that could not have found a place in hominid ancestry. But this line of argument appears to be based on quite arbitrary assumptions as to what should be regarded as specializations, and also on the false premise that even a slight degree of morphological specialization is not capable of undergoing an evolutionary reversal (see p. 46). So far as the canine tooth is concerned, it is first to be observed that in the Miocene and Pliocene genera of hominoids it actually showed a good deal of variation in size; we have already noted that in *Ramapithecus* it was unusually small. Second, there is reasonably good evidence of a direct anatomical nature that the small, spatulate canine of *H. sapiens* is the result of a secondary reduction in size. For example, the newly erupted and unworn canine (particularly the deciduous canine) may project markedly beyond the level of the adjacent teeth and may occasionally also be sharply pointed. Again, the permanent canine is provided with an unusually robust root, and the latter is also longer than that of the immediately adjacent teeth (fig. 34); such features are difficult to explain on a purely functional basis,

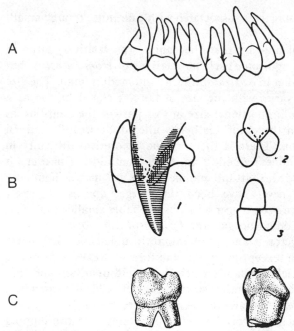

A

B

C

Fig. 34. Diagram illustrating A, the relatively long and robust root of the upper canine tooth in *Homo sapiens*; B, the occlusal relationships of the canines and a sectorial first lower premolar tooth in a cercopithecoid monkey (*1*), and the occlusal relationship of the canines and the first lower premolar tooth in *Homo sapiens* before attrition (*2*), and at an early stage of attrition (*3*) (after Remane); and C, lateral and anterior views of a first lower milk molar tooth of *Homo sapiens*.

for in modern man the canines have no special function. But they do become intelligible if we suppose that they had special functions in the past. Finally, the eruption of the permanent canine in *H. sapiens* is still late relative to the eruption of the adjacent teeth; it may come into place only after the two premolars and sometimes even after the second molar. There is also the evidence of fossil hominids that the modern canine has undergone retrogressive modifications, for (as we have noted) in *Homo erectus* the canines in some individuals were relatively large, overlapping teeth, associated with a definite, though small, diastema.

In review, it appears that the canine was relatively large in *Ramapithecus* and some early populations of *Homo erectus*, but somewhat smaller in *Australopithecus* and modern man. The size of the canine varies with the size of the anterior dentition as a whole, and the proportional size of this part of the dentition as compared with the cheek teeth has altered back and forth in man's evolution. There is therefore no theoretical difficulty in postulating canine reduction in early hominids. However, it should be added that the canine tooth of man's hominoid ancestor may never have been very large, and is extremely unlikely to have been bigger and was probably smaller than that seen in the Middle Miocene ape *Proconsul* (fig. 30).

That the sectorial and predominantly unicuspid first lower premolar tooth (except in its extreme form) is really a primitive and not a specialized feature of the hominoid dentition and that the bicuspid hominid tooth is secondarily derived is indicated by a number of considerations. In the first place, the sectorial type of tooth is a functional correlate of a projecting and overlapping upper canine, for it is shaped in conformity with the occlusal requirements of the latter. Second, in every genus of the earlier fossil Hominoidea so far known (even as far back as the little *Parapithecus* of Oligocene times), the first lower premolar is a predominantly unicuspid tooth,[2] and, indeed, this is a character of the primitive mammalian dentition as a whole. The evolutionary derivation of a bicuspid premolar from a unicuspid tooth is, after all, only one example of the tendency for an elaboration ("molarization") of the premolars that is a quite common feature in the phylogenesis of other groups of Primates and seen at its most extreme in *Australopithecus*. Last, as Remane (1927) has pointed out, in *H. sapiens* the morphology of the first deciduous molar, that is, the temporary precursor of the first lower premolar, offers very suggestive evidence, for it is often markedly compressed from side to side, with the anterolateral surface sloping obliquely downward, very much as that of a sectorial tooth does (fig. 34). It is not improbable that this feature is an example of the astonishing conservatism of morphological elements, as the result of which form and structure may persist for a long geological time after they have ceased to serve their original function.

These, then, are the lines of evidence that lead to the inference that the bicuspid premolar of hominid type is secondarily derived from a predominantly unicuspid and moderately sectorial tooth not very dissimilar to that still found in the modern large apes.[3] And it is to be noted that the inference accords well with the evidence that the brachydont hominid canine is a secondary modification of a conical, projecting canine. It hardly needs to be emphasized again that the final proof of such an evolutionary sequence must depend on adequate paleontological documentation; but it is at least clear that there is no theoretical reason why the hominid type of dentition should not have been derived from the generalized pongid type found in some of the Miocene apes, and such indirect evidence as is available certainly supports this proposition, as does the direct evidence of the Fort Ternan specimen.

Apart from the dentition, it seems that some authorities have been reluctant to accept the suggestion that the evolutionary precursors of the Hominidae could be represented by any of the known Miocene genera of apes, on the assumption that, because their dental characters show definitely that most of them were certainly pongids, they must therefore have shown aberrant specializations in limb structure, and so forth, comparable with those characteristic of the modern anthropoid apes. In fact, however, there has never been any real basis for such an assumption; on the contrary, by analogy with what is now known of the evolution of other mammalian groups, it might have been anticipated that the limb structure of these extinct hominoids would still have preserved a much more generalized condition— approximating in some degree, that is to say, that of the quadrupedal cercopithecoid monkeys. Paleontological discovery has shown that this certainly was the case, at least with some of the extinct genera. From Miocene deposits of Europe a femur[4] and the shaft of a humerus of *Dryopithecus* are known; there is also the skeleton of a small gibbonlike ape, *Pliopithecus*. Relatively complete, but rather fragmentary, remains of the limb skeleton of Miocene apes have been recovered from East Africa. Taken all together, this evidence indicates that the limb structure of these extinct apes had certainly not developed the extreme and aberrant specializations of the modern anthropoid apes. The shafts of the dryopithecine humerus and femur are slender and straight and suggest lightly built and agile creatures. The limb bones of *Pliopithecus* recently found in Middle Miocene deposits in Austria are, according to a most detailed study by Zapfe (1960), astonishingly primitive in their general construction and, in some of their morphological details, resemble those of cercopithecoid type. The humerus and femur of the Early Miocene genus *Proconsul* from East Africa are very similar to those of *Dryopithecus*, and the calcaneus and talus show interesting differences from those of the modern apes (Clark and Leakey 1951). In another East African Miocene pongid, *Limnopithecus*, which is evidently an extinct representative of the Hylobatinae (and perhaps not really generically distinct from *Pliopithecus*), the limb skeleton, in its proportions and morphological details, also shows a number of cercopithecoid characters (Clark and Thomas 1951). So far as it goes, then, the paleontological evidence suggests that in their limb structure at least some, and perhaps all, of the Miocene apes had not at that time developed the specializations related to the extreme form of brachiation shown by the modern apes, which is associated with characteristic changes in limb proportions. On the contrary, they were evidently in some ways more like the quadrupedal cercopithecoid monkeys of today, active and agile creatures, capable of scampering on the ground as well as leaping among the branches of the trees. The importance of these observations is clear, for they demonstrate that, so far as the proportions and structural adaptations of the limbs are concerned, there appears now to be no theoretical objection to the derivation of the Hominidae and the Recent Pongidae from a common ancestry at least as late as Early or Middle Miocene times. It was perhaps subsequent to this, therefore, that, in association with opposing trends in adaptation to posture and gait, the divergent evolutionary development of characteristic growth rates of the limb and trunk in the Hominoidea marked the initial phylogenetic separation of the earliest precursors of the Hominidae from the Pongidae.

It has been argued that, since the terms "anthropoid ape" and "pongid" were originally defined by reference to the modern anthropoid apes, which are characterized by certain specializations in limb proportions, and so forth, they can hardly be applied to extinct genera in which (as is now known) these specializations were not fully developed. But this is taking too static and narrow a view of the definition of larger taxonomic categories. In the case of the Miocene apes, their inclusion in the Pongidae appears valid for the following reasons: (1) a major taxonomic category should naturally include not only the terminal products of its evolution but also the earlier phases of development through which it passed after it became definitely segregated from immediately related groups; (2) it is to be expected that the earlier representatives of the Pongidae, after their initial segregation from the Cercopithecoidea, would retain primitive characters that became lost in the course of later evolution; (3) there is good evidence that the Cercopithecoidea had, in fact, already become a separate and well-defined group by the Early Miocene, for fossil remains of a cercopithecoid monkey, *Victoriapithecus*, have been recovered from the early Middle Miocene deposits of East Africa showing the specialized dental features distinctive of the whole group; (4) the dental morphology of the Miocene apes approximates very closely that of the Recent Pongidae, and in certain elements of the dentition (e.g., the canines and premolars) appears in some cases to be indistinguishable; (5) a difference in limb proportions does not by itself constitute an adequate basis for a familial distinction; (6) in spite of their general cercopithecoid appearance, the limb bones of some of the Miocene apes (e.g., *Limnopithecus*) do show certain features in which they approximate the Recent Pongidae. If all these factors are given due consideration, it seems reasonable to include these extinct genera of fossil apes (so far as they are at present known) in the Pongidae, though it may perhaps be justifiable to accord them a subfamilial distinction as Dryopithecinae.

The factors that determined the segregation of the evolutionary radiation of the Hominidae can be only conjectured. There is no reason to suppose that the Miocene apes were not arboreal (even though it seems likely that they were also capable of active progression on the ground), but there existed in the Early Miocene of East Africa a large ape, *Proconsul major*, which was evidently equivalent in size to the modern gorilla (Clark and Leakey 1951); this creature must presumably have been restricted in its arboreal habits to the main branches of the trees at the lower levels near the ground and may even have been mainly terrestrial. It would be particularly interesting to know something of the limb structure of *P. major*, but unfortunately no limb bones have been found.

That the adoption of bipedal terrestrial habits by the earliest representatives of the Hominidae must have occurred in regions of deforestation seems very probable; and it is perhaps significant, therefore, that the environment of the East African Miocene apes provided possibilities of this sort. A consideration of the evidence of the general faunal assemblage in the Miocene deposits of Kenya indicates that this environment consisted of wooded valleys of limited extent, separated by open savanna—there were probably no large continuous tracts of tropical forest comparable to those inhabited by the modern anthropoid apes in Central Africa. As we have seen, a reduction in the forested regions also occurred in Eurasia.

It may be suggested that the evolution of ground-living forms in the ancestry of the Hominidae was the result of adaptations primarily concerned not with the abandonment of arboreal life, but (paradoxically) with an attempt to retain it. For in regions undergoing gradual deforestation they would make it possible to cross intervening grasslands in order to pass from one restricted and shrinking wooded area to another. This proposition is parallel to the interesting conjecture that water-living vertebrates initially acquired terrestrial and air-breathing adaptations in order to preserve their aquatic mode of life; for in times of drought these adaptations would make it possible to escape from dried-up rivers or pools and go overland in search of water elsewhere.

In summary, the status of *Ramapithecus* is still somewhat controversial in the absence of postcranial material, but on the basis of the present evidence it seems entirely justifiable to classify the genus in the Hominidae. Our judgment on *Gigantopithecus* and *Sivapithecus*, as on *Oreopithecus*—all genera to which hominid status has been attributed in the past—must, however, remain negative in this respect.

It is hardly possible to prepare a very valuable definition for *Ramapithecus*, but our present knowledge can be summarized as follows:

> *Ramapithecus*—a genus of the Hominidae distinguished by the following characters: a small primate with short and deep face, with arched palate, maxillary zygoma originating above the first molar and somewhat flared; masticatory apparatus absolutely smaller and mandible shallower than in *Australopithecus*, but mechanics of jaw function as in other Hominidae: increased power of the cheek teeth for powerful crushing, reduced front teeth. Incisors and canines relatively small but less reduced than in *Australopithecus*; first lower premolar with small lingual cusp; molars low-crowned, steep-sided with broad, flat occlusal surfaces and simple cusp pattern; tightly packed, with thick enamel and a high rate of wear.

From this review of our three hominid genera, *Homo*, *Australopithecus*, and *Ramapithecus*, it is clear that the family itself cannot be defined by a static set of characters even if taken in combination, but must be defined by its dominant evolutionary trends[5] that have characterized this taxonomic group (and thereby serve to differentiate it from the evolutionary trends in opposite and contrasting directions that have characterized the Pongidae); and the formulation of these trends, again, must be based objectively on a consideration of the end-products of hominid evolution, as well as on the paleontological evidence so far available. For the paleontologist, also, the definition must be limited to those characters that are available for study in fossilized remains, that is, the skull, skeleton, and dentition. The following definition of the Hominidae is suggested:

> *Family Hominidae*—a subsidiary radiation of the Hominoidea distinguished from the Pongidae by the following evolutionary trends; progressive skeletal modifications in adaptation to erect bipedalism, shown particularly in a proportionate lengthening of the lower extremity, and changes in the proportions and morphological details of the pelvis, femur, and pedal skeleton related to mechanical requirements of erect posture and gait and to the muscular development associated therewith; preservation of well-developed pollex; ultimate loss of opposability of hallux. Increasing flexion of basicranial axis associated with increasing cranial height; relative displacement forward of the occipital condyles; restriction of nuchal area of occipital squama, associated with low position of inion; consistent and early ontogenetic development of a pyramidal mastoid process; reduction of subnasal prognathism, with ultimate early disappearance (by fusion) of facial component of premaxilla. Progressive alteration in relationship between cheek teeth and front teeth, the former enlarged for crushing and grinding with a powerful masticatory apparatus delivering increased direct power to the molar series; reduction in incisors; diminution of canines to a spatulate form, interlocking slightly or not at all and showing no pronounced sexual dimorphism; disappearance of diastemata; replacement of sectorial first lower premolars by bicuspid teeth (with later secondary reduction of lingual cusp); alteration in occlusal relationships, so that all the teeth tend to become worn down to a relatively flat, even surface at an early stage of attrition; development of an evenly rounded dental arcade; marked tendency in later stages of evolution to a reduction in size of the molar teeth; progressive acceleration in the replacement of deciduous teeth in relation to the eruption of permanent molars; progressive "molarization" of first deciduous molar. Marked and rapid expansion (in some of the terminal products of the hominid sequence of evolution) of the cranial capacity, associated with reduction in size of jaws and area of attachment of masticatory muscles and the development of a mental eminence.

It is to be noted that this provisional definition of the family Hominidae is not intended to be exhaustive, but merely to represent some of its main distinguishing features. It is also to be noted that all the characteristic evolutionary trends have not necessarily proceeded synchronously; as we have already emphasized, paleontological evidence of evolutionary sequences in general show that they not uncommonly proceed at different rates. But by an analysis of the total morphological pattern of a fossil hominoid (provided sufficient data are available) it should be possible, even in relatively early stages of their initial segregation and divergence from one another, to determine whether such a fossil is representative of the evolutionary sequence already committed by incipient changes to the developmental trends characteristic of the Hominidae, or to those characteristic of the Pongidae.

For comparison with the Hominidae, the Pongidae may be defined in the following terms.

Family Pongidae—a subsidiary radiation of the Hominoidea distinguished from the Hominidae by the following evolutionary trends: progressive skeletal modifications in adaptation to arboreal brachiation, shown particularly in a proportionate lengthening of the upper extremity as a whole and of its different segments; acquisition of a strong opposable hallux and modification of morphological details of limb bones for increased mobility and for the muscular developments related to brachiation; tendency to relative reduction of pollex; pelvis retaining the main proportions characteristic of quadrupedal mammals. Marked prognathism, with late retention of facial component of premaxilla and sloping symphysis; development (in larger species) of massive jaws associated with strong muscular ridges on the skull (see fig. 35); nuchal area of the occiput becoming extensive, with relatively high position of the inion; occipital condyles retaining a backward position well behind the level of the auditory apertures; only a limited degree of flexion of basicranial axis associated with maintenance of low cranial height; cranial capacity relative to body size showing no marked tendency to expansion; progressive hypertrophy of incisors with widening of symphyseal region of mandible and ultimate formation of "simian shelf"; enlargement of strong conical canines interlocking in diastemata and showing distinct sexual dimorphism; accentuated sectorialization of first lower premolar with development of strong anterior root; postcanine teeth preserving a parallel or slightly forward divergent alignment in relatively straight rows; first deciduous molar retaining a predominantly unicuspid form; no acceleration in eruption of permanent canine.

Again it is to be noted that these evolutionary trends have not all proceeded at the same rate, nor have they all been realized to the same degree at the same stage of evolution. For example,

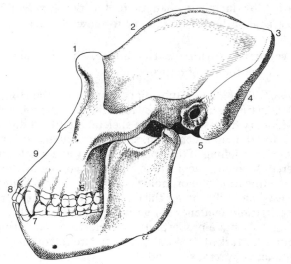

Fig. 35. Skull of male Gorilla. The characters numbered are: (*1*) The forward projection of the brow ridges well beyond the front end of the braincase; (*2*) the low level of the cranial roof in relation to the upper border of the orbital aperture; (*3*) the high position of the external occipital protuberance; (*4*) the steep slope and great extent of the nuchal area of the occiput for the attachment of the neck muscles; (*5*) the relatively backward position of the occipital condyles; (*6*) the teeth, which except in advanced stages of attrition are not worn down to a flat, even surface; (*7*) the canines, which form conical, projecting, and sharply pointed "tusks"; (*8*) the large size of the incisor teeth; (*9*) the massive upper jaw.

there is now good evidence that in the Miocene the dental morphology characteristic of the Recent Pongidae had already been acquired (except for the hypertrophy of the incisors and the associated development of a "simian shelf" in the mandible); but the limb skeleton at that time still retained many primitive features suggesting quadrupedal locomotion of the cercopithecoid type (Clark 1950; Clark and Leakey 1951; Clark and Thomas 1951). It seems probable, indeed, that the extreme specializations of the limbs for arboreal brachiation may have been a relatively late development in the evolution of the Pongidae.

Insofar as the term "Pongidae" refers to the subsidiary radiation of the Hominoidea that culminated in the development of the modern anthropoid apes, it must include all those types representative of the earlier phases of pongid evolution after this group had become definitely segregated from the Cercopithecoidea. Studies of comparative anatomy make it certain that the ancestral stock from which the modern anthropoid apes arose must have shown just such a combination of characters of an intermediate kind as those found in the Miocene hominoids; and a critical examination of the latter also makes it clear that they had already become definitely segregated from the Cercopithecoidea and that most of the known types exhibited the evolutionary trends characteristic of the Pongidae.

Those anatomists who have persuaded themselves of the "uniqueness" of man's anatomical structure have commonly assigned to him a vast geological antiquity. But, as has already been pointed out, there is no objective reason for assuming that the family of the Hominidae, *morphologically speaking*, is more "unique" than any other family of the Mammalia. So far as the evidence at present available can be assessed, the origin of the Hominidae and the Pongidae from a common ancestral stock seems well assured. There is no sound argument for pushing back the origin of the Hominidae to the Oligocene (or, as some would even suggest, to the Eocene). There is, in fact, no reason it may not have occurred in the Upper Miocene. It is here a question of attempting to estimate the time factor that is likely to have been involved in the evolutionary development of those characters that are distinctive of the Hominidae. But arguments based on the analogy of evolution rates in other mammalian groups are, unfortunately, far from secure, and those based on consideration of rates of molecular evolution, though suggestive, are without sound foundation. It is the discovery of fossil evidence that alone can give us a definitive answer to this question.

The interpretation of the paleontological evidence of hominid evolution that has been offered in the preceding chapters is a provisional interpretation. Because of the incompleteness of the evidence, it could hardly be otherwise. But it is an interpretation that appears to accord reasonably well with the facts now available. In one of his many writings, Karl Pearson made the observation, "Science consists not in absolute knowledge, but in the statement of the probable on the basis of our present—invariably limited—acquaintance with facts." To no branch of science can this be more aptly applied than to the science of paleoanthropology.

Notes

Chapter One

1. It is to be observed that mammalian classifications are not entirely of the "vertical" type. Where the fossil evidence is still wanting and in some cases where it is a matter of greater taxonomic convenience, terminal products of different lines of phylogenetic development may be grouped together in a "horizontal" type of classification. For example, in the Primates the Old World monkeys and the New World monkeys are at present grouped in a common suborder, the Anthropoidea, although there is sound evidence that these two groups of monkeys had their evolutionary origin in different groups of Eocene prosimians.

2. The Frankfurt plane and some of the other anatomical landmarks of the skull to which reference is made in the text are illustrated in figure 3, p. 35.

3. The fallacies that may arise from comparing overall measurements in types that are not closely related have also been emphasized by Washburn (1942).

4. Through the study of deep-sea cores it has been possible to establish a dated sequence showing (*a*) O^{16}/O^{18} ratio in the prehistoric ocean that indicates global ice volume; (*b*) radiolarian assemblage structure that indicates sea surface temperature: and (*c*) frequency of *Cycladophora*—another radiolarian—that indicates surface salinity and other structural aspects of the surface waters. This information, when combined, records both the extent and the chronology of world climatic changes. Up to the present these have been studied for the period since the beginning of the Middle Pleistocene, which is conventionally dated at 700,000 BP.

Chapter Two

1. A table showing the probable relationships of successive Paleolithic cultures to the glacial and interglacial periods of the Pleistocene is shown in figure 4, p. 54.

2. Pearson and Bell (1919), as a result of their biometric studies, concluded that the Galley Hill femur is distinct from that of Recent man. However, the Recent material on which they relied for their comparison was actually limited to a sample of the seventeenth-century London population, whereas, of course, "Recent man" must be assumed to include *all* modern races, as well as Neolithic man. This conclusion of Pearson and Bell's is an illustration of the fallacy of extrapolating from the special to the general in biometrics (see p. 7).

3. These dubious specimens include some skulls of modern type from Europe and North America, discovered when the sciences of stratigraphy and archeology were very poorly developed.

4. For a complete record of hominid fossil discoveries, the reader is referred to the *Catalogue of Fossil Hominids* prepared by Oakley, Campbell, and Molleson (1971–77). Where required, the numbered reference system for hominid fossils adopted in the *Catalogue* has been used in this book.

5. Some degree of taurodontism has been reported as an occasional variant in modern man, though rarely showing the extreme condition to be found in some representatives of later Mousterian man. Nevertheless, the character of taurodontism *by itself* is not definitely diagnostic of this group.

6. For an amplification of these morphological details see Boule and Vallois (1957), Howell (1951), Straus and Cave (1957).

7. In the original monograph on the Rhodesian remains (Pycraft et al. 1928), the innominate bone was held to be entirely anomalous in the form of its acetabular socket, and on the basis of this unusual feature a new genus, *Cyphanthropus*, was created. However, this conclusion was later shown to be the result of a rather remarkable error of interpretation (Clark 1928).

8. For a review on "New evidence of fossil man in China," see Chang 1962.

Chapter Three

1. It also remains possible, of course, that future discoveries will demonstrate that the owner of the *"Meganthropus"* mandible was so different from *H. erectus* in the morphological characters of the skull, dentition, and limb bones as to justify a generic distinction. But the point is that we are not per-

suaded that such a distinction is valid on the basis of the fragmentary material so far available.

2. Because of the space occupied by cerebrospinal fluids surrounding the brain and within its ventricles, it is important to realize that in man only two-thirds to three-fourths of the endocranial cavity is actually occupied by brain tissue (Mettler 1955).

3. Even in the young adult, S3, the parietal bone reaches a thickness of 10 mm.

4. This term is applied to the thin ledge of bone that, in the Recent large anthropoid apes, commonly extends back from the lower border of the symphysis across the anterior end of the interramal space of the mandible.

5. It is of particular interest that the overall dimensions of the first premolar and the first molar of the australopithecine maxilla found at Sterkfontein in 1936 differ from those of S4 by less than 0.5 mm.

6. Weidenreich's statement to the contrary seems to have been based on the comparison of *isolated* teeth and on the assumption that the baseline of the enamel of the canine tooth is level with that of the adjacent teeth. This, however, is by no means the case.

Chapter Four

1. *Australopithecus* means "the southern ape" and refers to the fact that this fossil hominoid lived in the Southern Hemisphere. The term has been criticized on etymological grounds. But its main disadvantage is that it is liable to convey to the reader of popular expositions the suggestion of a connection with Australia and also that (as it has now turned out) the creature is certainly not an anthropoid ape.

2. It is of interest to compare modern man, whose cranial capacity ranges (as we have seen) from ca. 1,000 to 2,200 cc (mean 1,350 cc); the chimpanzee, which varies between 282 and 500 cc (mean 383 cc); and the gorilla, which ranges between 340 and 752 cc (mean 505 cc) (Tobias 1971).

3. For a critical discussion of the endocranial casts of the Australopithecinae discovered up to 1947 see Clark (1947) and Holloway (1975).

4. This statement was controverted on the basis of observations that in a few gorilla skulls the index approximates closely that of the Sterkfontein skull. But these observations were based on measurements that most unfortunately included the sagittal crest in the cranial height of male gorillas (Clark 1952).

5. It should be noted that the degree of platycephaly present in *Homo erectus* lowers the vault and raises the index accordingly.

6. In the pygmy chimpanzee, *P. paniscus*, the condylar-position index may approximate closely that of the Sterkfontein skull, but the skull of this species preserves an infantile form and is small and very lightly constructed; it is thus not directly comparable with the more massive and heavily built australopithecine skull.

7. It has been assumed that the condylar-position index *by itself* is always correlated with the degree of postural erectness. The fallacy of this assumption is exposed by the fact that the index varies considerably even in modern *H. sapiens*. In extreme brachycephaly, for example, the occipital condyles may be placed considerably farther back in relation to the total skull length than in dolichocephalic skulls. But this can hardly be taken to mean that such brachycephalics do not habitually assume a fully erect posture!

8. It is a significant commentary on the close relationship of the Pongidae and the Hominidae that in some groups the molar teeth may be exceedingly difficult to distinguish with absolute certainty. The prolonged controversy on the molar teeth of the fraudulent Piltdown jaw (quite apart from the faked fabrication of the hominid type of wear) is sufficient to make this clear. But, as far as the australopithecine molars are concerned, though in the details of their cusp morphology they can be readily distinguished from all the known genera of apes, it would be extremely difficult to distinguish those in some populations from molars of *H. erectus*.

9. Campbell (1974) claims that the small glenoid angle adds further evidence to that from *H. sapiens* that some degree of brachiation occurred in the ancestry of the hominids.

10. For a recent consideration and rebuttal of this interpretation, see Howell, Washburn and Ciochon (1978).

11. Kromdraai carries one flake artifact in association with *Australopithecus robustus*: this is considered insufficient evidence on which to modify the preceding statement, especially considering the late age of the site and the known presence of more advanced hominids in the area (Sterkfontein and Swartkrans are within a mile). At Sterkfontein, Koobi Fora and Omo-Shungura early stone industries are not directly associated with *Australopithecus* but occur in later deposits.

12. Surface finds of quartz and lava artifacts are reported from the surface of older members (B, C, D, and E)—but none were unquestionably in situ (Merrick 1976).

Chapter Five

1. Two genera of small fossil hylobatids are also recognized from this period: *Limnopithecus* and *Pliopithecus* (see p. 196).

2. The statement has been made that, in the primitive hominoid mandible of *Propliopithecus* of Oligocene date, the premolars are homomorphic.

3. It has already been noted (p. 107) that in the genus *Homo* the lingual cusp of the first lower premolar (which is so conspicuous in *Australopithecus*) has undergone a secondary reduction.

4. This femur was discovered more than a hundred years ago, and for many years it was assumed to be that of a giant gibbon. But no remains of a giant gibbon have ever been found in Europe (or for that matter, elsewhere). The femur is certainly a pongid femur of some sort, and the only large Pongidae known to have existed contemporaneously in Europe are the dryopithecine apes. The femur conforms very well to the probable size of these apes, and there can therefore be little doubt that it must be referred to *Dryopithecus*.

5. It should perhaps be emphasized that the phrase "evolutionary trends" is not meant to refer here to the *inherent* trends of evolution that have been postulated by orthogeneticists. It refers to the graduated sequence of morphological changes that must obviously have occurred in phylogenetic history to produce the known end-products of evolution and that in some cases has been demonstrated (or at least partly confirmed) by paleontological evidence. Identical evolutionary trends in related groups imply a community of origin, since they must depend on the possession of similar genetic constitutions associated with similar potentialities for producing the same mutational variations, and on these, again, depends the ability to achieve the same adaptations. In other words, identical evolutionary trends imply phylogenetic relationship and are to be taken into account in assessing the homogeneity of major taxonomic groups. It may be argued that two independent groups derived from a remote ancestry might have followed identical evolutionary trends, leading to end-products not morphologically distinguishable. The answer to this argument is that, on the basis of the natural selection of random variations, the genetic probabilities are entirely against such a proposition, and in any case paleontology provides no evidence for extreme parallelism of this sort (see p. 18).

References

Aigner, J. S., and Laughlin, W. S. 1973. The dating of Lantian man and his significance for analysing trends in human evolution. *Am. J. Phys. Anthropol.* 39:97–109.

Andrews, P., and Tobien, H. 1977. New Miocene locality in Turkey with evidence on the origin of *Ramapithecus* and *Sivapithecus*. *Nature* 268:699–71.

Andrews, P., and Williams, D. B. 1973. The use of principal components analysis in physical anthropology. *Am. J. Phys. Anthropol.* 39:291–304.

Arambourg, C. 1957. Récentes découvertes de paléontologie humaine en Afrique du Nord française. In *Proceedings of the Third Pan-African Congress on Prehistory*, p. 186. London.

Barbour, G. F. 1949. Ape or man? *Ohio J. Sci.* 49:4.

Battaglia, R. 1948. Osso occipitale umano rinvenuto del giacimento Pleistocenico di Quinzano del Commune di Verona. *Palaeont. ital.* 42:1.

Berckhemer, F. 1933. Der Steinheimer Urmensch und die Tierwelt seines Lebensgebietes. *Naturw. Mschw. "Aus der Heimat."* 47:101–15.

Bergman, R. A. M., and Karsten, P. 1952. The fluorine content of *Pithecanthropus* and of other specimens from the Trinil fauna. *Proc. Kon. Akad. Wetensch., Amsterdam* 55:151.

Bilsborough, A. 1972. A multivariate study of evolutionary

change in the hominid cranial vault and some evolution rates. *J. Hum. Evol.* 2:387–403.

Black, D. 1930. On an adolescent skull of *Sinanthropus pekinensis. Palaeont. Sinica*, ser. D, 7:1.

———. 1933. On the endocranial cast of the adolescent *Sinanthropus* skull. *Proc. Roy. Soc. Lond.*, ser. B, 112:263.

Boaz, N. T., and Howell, F. C. 1977. A gracile hominid cranium from Upper Member G of the Shungura Formation, Ethiopia. *Am. J. Phys. Anthropol.* 46:93–108.

Boule, M., and Vallois, H. V. 1957. *Fossil men.* London: Thames and Hudson.

Bronowski, J., and Long, W. M. 1952. Statistics of discrimination in anthropology. *Am. J. Phys. Anthropol.* 10:385.

———. 1953. The australopithecine milk canines. *Nature* 172:251.

Broom, R., and Robinson, J. T. 1952. Swartkrans ape-man, *Paranthropus crassidens. Transvaal Mus. Mem.*, no. 6.

Broom, R.; Robinson, J. T.; and Schepers, G. W. H. 1950. The Sterkfontein ape-man, *Plesianthropus. Transvaal Mus. Mem.*, no. 4.

Broom, R., and Schepers, G. W. H. 1946. The South African fossil ape-men: The Australopithecinae. *Transvaal Mus. Mem.*, no. 2.

Brummelkamp, R. 1940. On the cephalization stage of *Pithecanthropus erectus* and *Sinanthropus pekinensis. Proc. Kon. Akad. Wetensch., Amsterdam* 43:3.

Butzer, K. W., and Isaac, G. L., eds. 1975. *After the Australopithecines: Stratigraphy, ecology and culture change in the Middle Pleistocene.* The Hague: Mouton.

Buxton, L. H. D. 1938. Platymeria and platycnemia. *J. Anat.* 73:31.

Buxton, L. H. D., and Morant, G. M. 1933. The essential craniological technique. *J. Roy. Anthropol. Inst.* 63:19.

Campbell, B. G. 1973. A new taxonomy of fossil man. *Yearbook Phys. Anthropol.* 17:194–201.

———. 1974. *Human evolution: An introduction to man's adaptations.* Chicago: Aldine.

Campbell, B. G., and Bernor, R. L. 1976. The origin of the Hominidae: Africa or Asia? *J. Hum. Evol.* 5:441–54.

Chang, K. C. 1962. New evidence of fossil man in China. *Science* 136:749.

Clark, J. Desmond. 1950. New studies on Rhodesian man. *J. Roy. Anthropol. Inst.* 77:13.

Clark, W. E. Le Gros. 1928. Rhodesian man. *Man* 28:206.

———. 1934. The asymmetry of the occipital region of the brain and skull. *Man* 34:1.

———. 1947. Observations on the anatomy of the fossil Australopithecinae. *J. Anat.* 81:300.

———. 1950. New palaeontological evidence bearing on the evolution of the Hominoidea. *Quart. J. Geol. Soc.* 105:225.

———. 1952. A note on certain cranial indices of the Sterkfontein skull no. 5. *Am. J. Phys. Anthropol.* 10:1.

———. 1954. *History of the Primates.* 4th ed. London: British Museum (Natural History). Reissued 1963, University of Chicago Press.

———. 1967. *Man-apes or ape-men?* New York: Holt, Rinehart and Winston.

———. 1971. *The antecedents of man.* 3d ed. Edinburgh: Edinburgh University Press.

Clark, W. E. Le Gros, and Leakey, L. S. B. 1951. *The Miocene Hominoidea of East Africa.* Fossil mammals of Africa, no. 1. London: British Museum (Natural History).

Clark, W. E. Le Gros, and Thomas, D. P. 1951. *Associated jaws and limb bones of Limnopithecus macinnesi.* Fossil mammals of Africa, no. 3. London: British Museum (Natural History).

Colbert, E. H. 1949. Some palaeontological principles significant in human evolution. In *Early man in the Far East*, p. 103. Studies in physical anthropology, no. 1. Philadelphia: American Association of Physical Anthropology. [Viking Fund].

Conroy, G. C., and Pilbeam, D. R. 1975. *Ramapithecus*: A review of its hominid status. In *Paleoanthropology: Morphology and Paleoecology* ed. R. H. Tuttle, pp. 59–86. The Hague: Mouton.

Cooke, H. B. S. 1952. Mammals, ape-men, and Stone Age men

in southern Africa. *S. African Archeol. Bull.* 7:59.

Coon, C. S. 1962. *The origin of races.* New York: Knopf.

Corrucini, R. S. 1974. Calvarial shape relationships between fossil hominids. *Yearbook Phys. Anthropol.* 18:89–109.

Dart, R. A. 1925. *Australopithecus africanus*: The man-ape of South Africa. *Nature* 115:195.

————. 1949. Innominate fragments of *Australopithecus prometheus. Nature* 7:301.

————. 1958. A further adolescent australopithecine ilium from Makapansgat. *Nature* 16:473.

Day, M. H. 1969. Omo human skeletal remains. *Nature* 222: 1135–38.

————. 1971. Postcranial remains of *Homo erectus* from bed IV, Olduvai Gorge, Tanzania. *Nature* 232:383–87.

————. 1976. Hominid postcranial material from bed I, Olduvai Gorge. In *Human origins*, ed. G. L. Isaac and E. R. McCown. Menlo Park, Calif.: Benjamin.

Day, M. H., and Molleson, T. I. 1973. The Trinil femora. *Symposia for the Society for the Study of Human Biology* 11:127–54.

Dubois, E. 1932. The distinct organization of *Pithecanthropus erectus* now confirmed from other individuals of the described species. *Proc. Kon. Akad. Wetensch., Amsterdam* 35:716.

Eckhart, R. B. 1975. *Gigantopithecus* as a hominid. In *Paleoanthropology: Morphology and Paleoecology*, ed. R. H. Tuttle, pp. 105–29. The Hague: Mouton.

Elliot Smith, G. 1931. *The search for man's ancestors.* London: Watts.

Ennouchi, E., 1962. Un neandertalien: L'homme du Jebel Irhoud (Maroc). *Anthropologie* 66:279–99.

Ford, E. B. 1938. The genetic basis of adaptation. In *Evolution*, ed. G. R. de Beer, p. 43. London: Oxford University Press.

Frayer, D. W. 1974*a*. *Gigantopithecus* and its relationship to *Australopithecus. Am. J. Phys. Anthropol.* 39:413–26.

————. 1974*b*. A reappraisal of *Ramapithecus. Yearbook Phys. Anthropol.* 18:19–30.

Garn, S. M. 1971. *Human races.* 3d ed. Springfield, Ill.: Charles C. Thomas.

Gorjanovič-Kramberger, K. 1906. *Der diluviale Mensch von Krapina in Kroatien.* Wiesbaden: C. W. Kriedler.

Gregory, W. K., and Hellman, M. 1923. Notes on the type of *Hesperopithecus haroldcookii. Am. Mus. Novitiates*, no. 53, p. 1.

Gregory, W. K.; Hellman, M.; and Lewis, G. E. 1938. *Fossil anthropoids of the Yale-Cambridge Indian Expedition of 1935.* Carnegie Institution of Washington Publications, no. 495. Washington, D.C.: Carnegie Institution.

Haldane, J. B. S. 1949. Suggestions as to quantitative measurement of rates of evolution. *Evolution* 3:51.

Haxton, H. A. 1947. Muscles of the pelvic limb: A study of the differences between bipeds and quadrupeds. *Anat. Rec.* 98:337.

Hays, J.; Imbrie, J.; and Shackleton, N. J. 1977. Variations in the earth's orbit: Pacemaker of the ice-ages. *Science* 194:1121–32.

Heiple, K. G., and Lovejoy, C. D. 1971. The distal femoral anatomy of *Australopithecus. Am. J. Phys. Anthropol.* 35: 75–84.

Hirschler, P. 1942. *Anthropoid and human endocranial casts.* Amsterdam: N. V. Noord-Hollandsche Uitgevers Maatschapij.

Holloway, R. L. 1975. Early hominid endocasts: Volumes, morphology, and significance for hominid evolution. In *Primate functional morphology and evolution*, ed. R. H. Tuttle. The Hague: Mouton.

Hooijer, D. A. 1951. The geological age of *Pithecanthropus, Meganthropus*, and *Gigantopithecus. Am. J. Phys. Anthropol.* 9:265.

Howell, F. C. 1951. The place of Neanderthal man in human evolution. *Am. J. Phys. Anthropol.* 9:379.

———. 1952. Pleistocene glacial ecology and the evolution of "classic Neanderthal" man. *Southwestern J. Anthropol.* 8:377.

Howell, F. C., and Coppens, Y. 1974. Inventory of remains of Hominidae from Pliocene/Pleistocene formations of the Lower Omo Basin, Ethiopia (1967–1972). *Am. J. Phys. Anthropol.* 40:1–16.

Howell, F. C.; Washburn, S. L.; and Ciochon, R. L. 1978.

216
References

Relationship of *Australopithecus* and *Homo. J. Hum. Evol.* 7:127–31.

Howells, W. W. 1975. Neanderthal man: Facts and figures. In *Paleoanthropology: Morphology and paleoecology*, ed. R. H. Tuttle, pp. 389–407. The Hague: Mouton.

Hughes, A. R., and Tobias, P. V. 1977. A fossil skull probably of the genus *Homo* from Sterkfontein, Transvaal. *Nature* 265: 310–12.

Hürzeler, J. 1958. *Oreopithecus bambolii. Verhandl. Natur. Gesell. Basel* 69:1.

Jacob, T. 1973. Palaeoanthropological discoveries in Indonesia with special reference to the finds of the last two decades. *J. Hum. Evol.* 2:473–85.

———. 1975. Morphology and paleoecology of early man in Java. In *Paleoanthropology: Morphology and paleoecology*, ed. R. H. Tuttle, pp. 311–25. The Hague: Mouton.

Jaeger, J.-J. 1975. The mammalian faunas and hominid fossils of the Middle Pleistocene of the Maghreb. In *After the Australopithecines*, ed. K. Butzer et al., pp. 399–418. The Hague: Mouton.

Johanson, D. C., and Taieb, M. 1976. Plio-Pleistocene hominid discoveries in Hadar, Ethiopia. *Nature* 260:293–97.

Keith, A. 1931. *Further discoveries relating to the antiquity of man*. London: Williams and Norgate.

Koenigswald, G. H. R. von. 1936. Erste Mitteilung ueber einen fossilen Hominiden aus dem Altpleistocän Ostjavas. *Proc. Kon. Akad. Wetensch., Amsterdam* 39:1.

———. 1937. Ein Unterkieferfragment des *Pithecanthropus* aus den Trinilschichten Mitteljavas. *Proc. Kon. Akad. Wetensch., Amsterdam* 40:1.

———. 1938. Ein neuer *Pithecanthropus*-Schädel. *Proc. Kon. Akad. Wetensch., Amsterdam* 41:1.

———. 1940. Neue *Pithecanthropus*-Funde 1936–1938. *Wetensch. Meded.*, no. 28.

———. 1949. The discovery of early man in Java and southern China. In *Early man in the Far East*, p. 83. Studies in physical anthropology, no. 1. Philadelphia: American Association of Physical Anthropology.

————. 1950. Fossil hominids from the Lower Pleistocene of Java. *Proceedings of the Eighteenth International Geological Congress*, part 9, p. 59.

Koenigswald, G. H. R. von, and Weidenreich, R. 1939. The relationship between *Pithecanthropus* and *Sinanthropus*. *Nature* 184:491.

Kretzoi, M. 1975. New ramapithecines and *Pliopithecus* from the Lower Pliocene of Rudabanya in north-eastern Hungary. *Nature* 257:578–81.

Leakey, L. S. B. 1959. A new fossil skull from Olduvai. *Nature* 184:491.

————. 1961. New finds at Olduvai Gorge. *Nature* 189:649.

Leakey, L. S. B.; Tobias, P. V.; and Napier, J. R. 1964. A new species of the genus *Homo* from Olduvai Gorge. *Nature* 202:7–9.

Leakey, M. D. 1971. *Olduvai Gorge*. Vol. 3. *Excavations in Beds I and II, 1960–1963*. Cambridge: Cambridge University Press.

Leakey, R. E. F. 1973. Evidence for an advanced Plio-Pleistocene hominid from East Rudolf, Kenya. *Nature* 242:447–50.

————. 1976. New hominid fossils from the Koobi Fora Formation in northern Kenya. *Nature* 261:574–76.

Leakey, R. E. F., and Leakey, M. G., eds. 1978. *Koobi Fora: Researches into geology, palaeontology and human origins*. Vol. 1. Oxford: Oxford University Press.

Leakey, R. E. F., and Walker, A. C. 1976. *Australopithecus, Homo erectus* and the single species hypothesis. *Nature* 261:572–74.

Lewis, G. E. 1934. Preliminary notice of new man-like apes from India. *Am. J. Sc.* 27:161.

Lovejoy, C. O. 1978. A biomechanical review of the locomotor diversity of early hominids. In *Early hominids of Africa*, ed. C. Jolly. London: Duckworth.

Lovejoy, C. O.; Heiple, K. G.; and Burstein, A. H. 1973. The gait of *Australopithecus*. *Am. J. Phys. Anthropol.* 38:757–80.

Lumley, H. de, and Lumley, M. A. de. 1973. Pre-Neanderthal human remains from Arago Cave in southeastern France. *Yearbook Phys. Anthropol.* 17:162–68.

McCown, T. D., and Keith, A. 1939. *The Stone Age of Mount Carmel.* Oxford: Clarendon Press.

McHenry, H. M. 1975. A new pelvic fragment from Swartkrans and the relationship between robust and gracile australopithecines. *Am. J. Phys. Anthropol.* 43:245–61.

McHenry, H. M., and Corrucini, R. S. 1975. Distal humerus in hominoid evolution. *Folia Primat.* 23:227–44.

McHenry, H. M.; Corrucini, R. S.; and Howell, F. C. 1976. Analysis of an early hominid ulna from the Omo Basin, Ethiopia. *Am. J. Phys. Anthropol.* 44:295–304.

Mann, A., and Trinkaus, E. 1973. Neandertal and Neandertal-like fossils from the Upper Pleistocene. *Yearbook Phys. Anthropol.* 17:169–93.

Mayr, E. 1942. *Systematics and the origin of species.* New York: Columbia University Press.

————. 1970. *Populations, species, and evolution.* Cambridge: Harvard University Press.

Merrick, H. V. 1976. Recent archaeological research in the Plio-Pleistocene deposits of the lower Omo, southwestern Ethiopia. In *Human origins*, ed. G. L. Isaac and E. R. McCown, pp. 461–82. Menlo Park, Calif.: Benjamin.

Mettler, F. A. 1955. *Culture and the structural evolution of the nervous system.* New York: American Museum of Natural History.

Morant, G. M. 1927. Studies of Palaeolithic man. II. A biometric study of Neanderthaloid skulls and their relationships to modern racial types. *Ann. Eugenics* 2:318.

————. 1928. Studies of Palaeolithic man. III. The Rhodesian skull and its relations to Neanderthaloid and modern types. *Ann. Eugenics* 3:337.

Napier, J. R. 1959. *Fossil metacarpals from Swartkrans.* Fossil mammals of Africa, no. 17. London: British Museum (Natural History).

Oakley, K. P. 1950. New studies on Rhodesian man. *J. Roy. Anthropol. Inst.* 77:7.

———. 1951. A definition of man. *Sci. News* 20:69.

———. 1953. Dating fossil human remains. In *Anthropology today*, ed. A. L. Kroeber, p. 43. Chicago: University of Chicago Press.

———. 1964a. *Man the tool-maker.* 3d ed. Chicago: University of Chicago Press.

———. 1964b. *The problem of man's antiquity.* London: British Museum (Natural History).

———. 1969. *Frameworks for dating fossil man.* 3d ed. Chicago: Aldine.

Oakley, K. P.; Campbell, B. G.; and Molleson, T. I. 1971–77. *Catalogue of fossil hominids.* 3 vols. London: British Museum (Natural History).

Oakley, K. P., and Montagu, M. F. A. 1949. A reconsideration of the Galley Hill skeleton. *Bull. Brit. Mus. (Nat. Hist.) Geol. Ser.* 2:46.

Oppenoorth, W. F. F. 1932. *Homo (Javanthropus) soloensis*: Een Plistoceene mensch von Java. *Wetensch. Meded. Dienst Mijnbouw Nederlandsch-Indië* 20:49.

Ovey, C. D., ed. 1964. *The Swanscombe skull.* London: Royal Anthropological Institute.

Pearson, K., and Bell, J. 1919. *A study of the long bones of the English skeleton.* Part 1, sec. 2. Drapers Company research mem., biometric series, vol. 11. London: Draper's Company.

Pilbeam, D. R., and Gould, S. J. 1974. Size and scaling in human evolution. *Science* 186:892–901.

Pycraft, W. P., et al. 1928. *Rhodesian man and associated remains.* London: British Museum (Natural History).

Remane, A. 1922. Beiträge zur Morphologie des Anthropoidegebisses. *Arch. Naturgesch.* 87:1.

———. 1927. Studien ueber die Phylogenie des menschlichen Eckzahnes. *Ztschr. Anat. Entwicklungsgesch.* 82:391.

Robinson, J. T. 1954. The genera and species of the Australopithecinae. *Am. J. Phys. Anthropol.* 12:181.

―――. 1956. The dentition of the Australopithecinae. *Transvaal Mus. Mem.*, no. 9.

―――. 1972. *Early hominid posture and locomotion.* Chicago: University of Chicago Press.

Rosen, S. I., and McKern, T. W. 1971. Several cranial indices and their relevance to fossil man. *Am. J. Phys. Anthropol.* 35:69–74.

Schultz, A. H. 1931. Man as a Primate. *Sci. Monthly* 33:385.

―――. 1935. Eruption and decay of the permanent teeth in Primates. *Am. J. Phys. Anthropol.* 19:489.

―――. 1936. Characters common to higher Primates and characters specific for man. *Quart. Rev. Biol.* 11:259, 425.

―――. 1950*a*. *The anatomy of the gorilla.* New York: Columbia University Press.

―――. 1950*b*. The physical distinctions of man. *Proc. Am. Phil. Soc.* 94:428.

Sergi, S. 1943. Craniometria e iconografia del secondo Paleantropo di Saccopastore. *Rend. Accad. Italia* 7:41.

―――. 1944. Craniometria e craniografia del primo Paleantropo di Saccopastore. In *Estratto delle ricerche di morfologia*, p. 1 Rome: Istituto di Antropologia della Università di Roma.

Shellshear, J., and Elliot Smith, G. 1934. A comparative study of the endocranial cast of *Sinanthropus*. *Phil. Trans. Roy. Soc.*, ser. B, 223:469.

Simons, E. L. 1961. *The phyletic position of Ramapithecus*, Postilla no. 57. New Haven: Peabody Museum (Yale University).

―――. 1963. Some fallacies in the study of hominid phylogeny. *Science* 141:879.

―――. 1972. *Primate evolution: An introduction to man's place in nature.* New York: Macmillan.

―――. 1977. *Ramapithecus. Sci. Am.* 236:28–35.

Simons, E. L., and Pilbeam, D. R. 1965. Preliminary revision of the Dryopithecinae (Pongidae, Anthropoidea). *Folia Primat.* 3:81–152.

Simpson, G. G. 1945. The principles of classification and a classification of mammals. *Bull. Am. Mus. Nat. Hist.*, vol. 85.

————. 1950*a*. *The meaning of evolution.* London: Oxford University Press.

————. 1950*b*. Some principles of historical biology bearing on human origins. *Cold Spring Harbor Symp. Quant. Biol.* 15:55.

————. 1951. *Horses.* New York: Oxford University Press.

————. 1961. *Principles of animal taxonomy.* New York: Columbia University Press.

Singer, R. 1954. The Saldanha skull from Hopefield, South Africa. *Am. J. Phys. Anthropol.* 12:345.

Solecki, R. S. 1960. Three adult Neanterthal skeletons from Shanidar cave in northern Iraq. *Smithsonian Rep. Pub.*, no. 4414, p. 603.

Stewart, T. D. 1960. Form of the pubic bone in Neanderthal man. *Science* 131:1437.

Straus, W. L. 1949. The riddle of man's ancestry. *Quart. Rev. Biol.* 24:200.

Straus, W. L., and Cave, A. J. E. 1957. Pathology and posture of Neandertal man. *Quart. Rev. Biol.* 32:348–63.

Stringer, C. B. 1974*a*. A multivariate study of the Petralona skull. *J. Hum. Evol.* 3:397–404.

————. 1974*b*. Population relationships of Later Pleistocene hominids: A multivariate study of available crania. *J. Archaeol. Sci.* 1:317–42.

Thoma, A. 1966. L'occipital de l'homme Mindelien de Vértesszöllös. *L'Anthropologie, Paris* 70:495–534.

————. 1969. Biometrische Studie über das Occipitale von Vértesszöllös. *Z. Morph. Anthropol.* 60:229–41.

Tobias, P. V. 1967. *Olduvai Gorge.* Vol. 2. *The cranium of Australopithecus (Zinjanthropus) boisei.* Cambridge: Cambridge University Press.

————. 1971. *The brain in hominid evolution.* New York: Columbia University Press.

Tobias, P. V., and Koenigswald, G. H. R. von. 1965. Comparison between the Olduvai hominines and those of Java and some

implications for hominid phylogeny. *Nature* 204:515-18.
Trinkaus, E. 1975. Squatting among the Neandertals. *J. Archaeol. Sci.* 2:327-51.
————. 1976. The morphology of European and south west Asian Neandertal pubic bones. *Am. J. Phys. Anthropol.* 44: 95-104.
Twiesselman, F. 1961. Le fémur néanderthalien de Ford-de-Forêt. *Mem. Inst. Roy. Sci. Nat. Belgique* 148:1.
Vrba, E. S. 1975. Some evidence of chronology and palaeoecology of Sterkfontein, Swartkrans and Kromdraai from the fossil Bovidae. *Nature* 254:301-4.
Walker, A., and Andrews, P. 1973. Reconstruction of the dental arcades of *Ramapithecus wickeri*. *Nature* 244:313-14.
Washburn, S. L. 1942. Technique in primatology. *Anthropol. Briefs*, no. 1, p. 6.
————. 1950. The analysis of Primate evolution, with particular reference to the origin of man. *Cold Spring Harbor Symp. Quant. Biol.* 15:67.
Watson, D. M. S. 1951. *Palaeontology and modern biology*. New Haven: Yale University Press.
Weidenreich, R. 1928. *Der Schädelfund von Weimar-Ehringsdorf*. Jena: Gustav Fischer.
————. 1936. The mandibles of *Sinanthropus pekinensis*. *Paleont. Sinica*, ser. D, vol. 7:1-162.
————. 1937. The dentition of *Sinanthropus pekinensis*. *Palaeont. Sinica*, new ser. D, vol. 1.
————. 1939. On the earliest representatives of modern mankind recovered on the soil of East Asia. *Pekin Nat. Hist. Bull.* 13:161.
————. 1941. The extremity bones of *Sinanthropus pekinensis*. *Palaeont. Sinica*, no. 116.
————. 1943. The skull of *Sinanthropus pekinensis*. *Palaeont. Sinica*, no. 127.
————. 1946. *Apes, giants, and man*. Chicago: University of Chicago Press.
————. 1951. Morphology of Solo man. *Anthropol. Papers Am. Mus. Nat. Hist.* 43:205.

Weinert, H. 1936. Der Urmenschenschädel von Steinheim. *Ztschr. Morphol. Anthropol.* 35:463.

————. 1951. Ueber die neuen Vor- und Frühmenschenfunde aus Afrika, Java, China und Frankreich. *Ztschr. Morphol. Anthropol.* 42:113.

Wells, L. H. 1950. New studies on Rhodesian man. *J. Roy. Anthropol. Inst.* 77:11.

White, T. D. 1977. New fossil hominids from Laetolil, Tanzania. *Am. J. Phys. Anthropol.* 46:197–216.

Zapfe, H. 1960. Die Primatenfunde aus der miozänen Spaltenfüllung von Neudorf. *Schweiz. Paleont. Abhandl.* 78:1.

Zeuner, F. E. 1940. *The age of Neanderthal man.* Occasional papers of the Institute of Archaeology, University of London, no. 3. London: Institute of Archaeology.

Index

RANGE   MEANS

A   420 - 530   A.R. 500 - 530
H H   600 - 750   670
N E   SH 915 - 1025   1043
            S   900
H S - N   1300 - 1800   1350
           1000 - 2000

P T   883
P P   342
GONEN   --- 752